# Actively Caring for People's Safety

How to Cultivate a

Brother's/Sister's Keeper Work Culture

# Actively Caring for People's Safety

## How to Cultivate a Brother's/Sister's Keeper Work Culture

**E. Scott Geller, Ph.D.**
**and**
**Krista S. Geller, Ph.D.**

American Society of Safety Engineers
PARK RIDGE, ILLINOIS

American Society of Safety Engineers, Park Ridge, IL 60068


ISBN-13: 978-0-939874-10-1



A catalog record for this book is available from the Library of Congress.

Text design and composition: Cathy Lombardi
Cover design: Troy Courson, Image Graphics
Project and managing editor: Rick Blanchette, ASSE

Printed in the United States of America

24 23 22 21 20 19 18 17     1 2 3 4 5 6 7 8 9

# CONTENTS

# FOREWORDS

I've had the privilege of sitting by the fireplace with Scott Geller and his daughter, Krista, at the Make-A-Diff Ranch Lodge late into the night, discussing safety and how to prevent injuries. Scott is one of the most self-actualized (his ideals align with his behavior) and competent persons I have ever met. He educates others most effectively and passionately in the complex field of applied behavioral science. Many academics have profound knowledge, but they do not have his ability to share that expertise with the public at large. It's exciting that Dr. Geller's enthusiasm for helping others inspired his daughter (also Dr. Geller) to pursue the same passion for making a difference in the lives of others.

With this book, you have the opportunity to learn from the creator of Actively Caring for People (AC4P). Scott and his coauthor, Krista, have written a book relevant for the newest employee as well as the seasoned safety professional or CEO. Wisdom is learning from the experience of others and discovering what works and what does not work to make a beneficial difference. Scott has taken what has been learned from behavior-based safety (BBS), which he created in 1979, and then refined and improved it into AC4P.

This book has a tool I suggest you check out after you read the preface. Go to the Glossary of Key Terms on page 147. Take a few moments before reading this book and read through the glossary. This will familiarize you with the unique and critical language used throughout this teaching/learning guidebook. Also, it is a great place to refer to if you need to review a term or word.

You will learn how to share AC4P with others, as well as to understand the foundational principles. You will believe in these research-based concepts and techniques and feel empowered to apply them to the unique situations in your workplace. Understanding the "why" behind AC4P makes it possible for you to motivate others to join what is truly an international movement. I have coached many professionals—including Scott—on the most effective presentation skills, and I know that when I teach basic truths my clients can use them in varied situations.

AC4P is all about taking evidence-based behavioral-science principles, merging them with elements of humanism (Maslow's hierarchy of needs, for example) or what Scott calls "humanistic behaviorism" and creating a powerful new paradigm for workplace safety excellence. In this book, you will go beyond the common BBS training and learn new concepts, such as "feedforward." Everyone is familiar with feedback, but what do you do when you can't deliver observational feedback in a timely fashion? The answer is feedforward, as Drs. Geller explain and illustrate in Lesson 3.

Whether you are studying this profound knowledge alone or in a group setting, Scott and Krista have placed exercises and discussion questions throughout the book to help you apply immediately what you are learning. It is like having them in the room with you, coaching you on how to make the process your own.

One of the biggest challenges to both new and experienced safety professionals is digging through the many behavioral safety myths and common-sense solutions, which are not based on objective research. For instance, consider the phrase "praise publicly and reprimand privately." The authors point out the need to get permission before praising publicly, but it's best to praise privately in a one-to-one personal conversation.

Scott and Krista's book brings you peer-reviewed and fully researched techniques you can count on. When you learn and practice what they teach you, you will most certainly be effective at improving safety-related behavior, preventing personal injury to yourself and others, and promoting human welfare and well-being.

I am excited you are able to learn from these two dedicated professional teachers who truly care for others by providing scholarship and practical techniques that have reached people and organizations worldwide. I know many children will have Mom and Dad come home safely from work every day because of the principles and procedures Scott and Krista teach. Enjoy this book and the positive results you will surely experience.

John Drebinger, CSP
Certified Speaking Professional
President, John Drebinger Presentations

As I was completing a master's degree in industrial/organizational psychology in 1989 at West Chester University, most of my fellow graduates were heading directly to the job market. However, I had become fascinated with the behavioral science theory and methods I had learned, and I asked my professor at the time where I could go to learn more about applied behavioral science targeted towards industry. His first response was Scott Geller at Virginia Tech.

Thus, I moved on to Virginia Tech, first for another master's degree and then earning my Ph.D. under the guidance of Dr. Geller. Upon arriving at Virginia Tech, I began focusing my interests on industrial safety. It was here, in the early 1990s, I have my first memory of the term "Actively Caring." A group of Exxon Chemical executives were visiting to learn more about the human dynamics of safety. There was a discussion between Scott, a group of these executives, and me. One of them complained about the safety culture and said, "No one cares about safety."

Another said, "That's not true, everyone cares about safety."

A third spoke up: "I think you're both right—everyone cares about safety, we just have too few actually doing something about it."

"What you mean is we need more actively caring," said Scott.

Ever since, Scott, myself, and our consulting partners at Safety Performance Solutions have been using the AC4P principles described in this book to help both government and private organizations across practically every industry type achieve world-class safety performance. Following the principles described in this book will allow you to move beyond mere behavior-based safety to fully integrate the whole person and relevant context when understanding, intervening, and maintaining safe behavior in ways that encourage involvement, ownership, and empowerment at all levels of the organization.

Steve Roberts, Ph.D.
Senior Partner
Safety Performance Solutions, Inc.

# PREFACE

For more than three decades, behavior-based safety (BBS) has been flourishing in industrial settings across the U.S., and more recently throughout the world. Several books detail the principles and procedures of BBS.[1] Moreover, a number of systematic reviews of the literature provide solid evidence for the success of this approach to injury prevention.[2] None of these resources provide a practical teaching/learning guide to undertaking the research-based principles of BBS for real-world application. This is the mission of this teaching/learning manual. As explained later, we now refer to the BBS process as actively caring for people (AC4P).

Safety professionals readily relate to the term *behavior* because they realize human behavior is involved to some extent in almost every environmental hazard, personal injury, and fatality. So, improving safety-related behavior is essential for injury prevention. In other words, safety leaders and consultants recognize that unsafe (or at-risk) behavior contributes to almost every minor and major injury and fatality in the workplace. Therefore, substituting safe behavior for at-risk behavior is crucial for an upstream proactive approach to industrial health and safety.

Unfortunately, there has been much distortion and confusion among safety professionals and consultants regarding BBS. There's much more to BBS than believing behavior change is critical for injury prevention. This teaching/learning handbook reviews the theoretical and empirical foundation of BBS and shows you how to apply the best of BBS to keep people safe.

## Dream Chasing with Applied Behavioral Science (ABS)

In his presidential address to the Association for Behavior Analysis International in 2013, Dr. Richard Malott gave me (Scott Geller) the label Dream

Chaser. After helping to implement BBS for Ford Motor Company in the early 1980s, my dream chasing to make a difference was aided by attending regional and national safety conferences sponsored by the National Safety Council (NSC) and the American Society of Safety Engineers (ASSE). Frankly, I was astounded and disappointed by the myriad of unfounded presentations that attempted to address the human dynamics of injury prevention. It took significant self-control to keep me from leaping from my chair and shouting out the numerous fallacies.

What I heard time and again was an emphasis on top-down enforcement with *safety-cop* language. Psychology informs us that this traditional approach to occupational safety limits the employee engagement needed to cultivate an injury-free workplace. The common focus on safety attitudes and safety thinking was nebulous and a waste of time without first addressing safe versus at-risk behaviors.

Furthermore, we know from research (and common sense) that the typical safety incentive program that bases a prize or financial reward on avoiding injuries (actually, the reporting of injuries) encourages a failure-avoiding mind-set. Minor injuries go unreported, and the interpersonal conversations needed to prevent injuries are stifled. As depicted in the illustration below, a safety incentive program based on injury rate does not motivate the occurrence of safety-related behavior.

Dale Gray, corporate safety director of Ford Motor Company, put me in a position to voice my concerns to large audiences. Dale, a fellow of the American Society of Safety Engineers (ASSE), got me on the national conference program for ASSE in 1987, and since then I've offered a presentation related to the psychology of safety at the ASSE annual conference almost every year. Plus, I've given a BBS workshop and keynote address at the National Safety Council's annual National Safety Congress and Expo every year since 1995.

After hearing my 1990 ASSE address, Dave Johnson, the editor of *Industrial Safety and Hygiene News* (*ISHN*), facilitated my dream chasing further when he invited me to write for his safety magazine, which has more than 75,000 subscribers. This led to me writing an *ISHN* article every month for nineteen consecutive years. Each month my *Psychology of Safety* column connected an evidence-based principle and/or research finding in psychology with a particular safety-related issue or challenge. This monthly assignment challenged me to explore the variety of human dynamics related to keeping people safe, and it enabled me to evolve as both a researcher and a communicator of behavioral and psychological science.

## Safety Performance Solutions, Inc.

In 1991, five young professionals, including two of my former Ph.D. students and an M.S. student, incorporated a partnership with me to help companies customize principles of ABS for injury prevention. This involved education, training, coaching, and consulting. Initially, we called our approach behavior-based safety (BBS); later we used the term people-based safety (PBS); and today we use the more general term: actively caring for people (AC4P).

The partners and affiliates of Safety Performance Solutions (SPS) have traveled far and wide to address occupational safety with applied behavioral science (ABS). From teaching the basics of BBS, the SPS toolbox has expanded to include the administration and scoring (with extrinsic benchmark comparisons) of a comprehensive research-based culture survey, as well as a unique 360-degree feedback and interpersonal coaching approach to helping company leaders improve their beneficial impact on the human dynamics of safety and the organization's bottom line.

Almost every year since its inception, SPS has hosted an AC4P User's Conference, which brings together safety professionals, corporate leaders, and wage workers for the sharing of various practical applications of AC4P principles to keep people safe. Keynote presentations and workshops are not given exclusively by SPS partners—at least 50% of the presentations are given by actual "users" of ABS technology. They discuss creative applications and behavioral results, as well as barriers to obtaining corporate-wide buy-in and strategies for overcoming them. Watching these diverse corporate folks discuss ways to bring ABS principles to life to successfully prevent workplace injuries and fatalities has been incredibly inspirational.

## A Proud Dad

I am extremely proud to coauthor this teaching/learning guide with my oldest daughter, Krista S. Geller, Ph.D. Throughout her childhood, Krista showed sincere interest in the ABS research my students and I were conducting to develop, evaluate, and refine interventions for improving behavior related to human welfare and quality of life. For example, at age sixteen, she and her fourteen-year-old sister helped me demonstrate that it was too easy for underage teenagers to purchase cigarettes in our university town. Each daughter was equipped with a miniature tape recorder, and then my graduate students drove them to twenty different stores to purchase a pack of cigarettes. If a cashier asked for identification to prove they were eighteen or older, they were coached to say, "Oh no! I left my driver's license at home, but please let me buy this one pack of cigarettes."

Fourteen-year-old Karly was turned down only twice, once when her basketball coach was the cashier, and sixteen-year-old Krista was never turned down. When I reported these discouraging findings to the local newspaper, our field study became front-page news. The headline read, "The Sting in Blacksburg," and the article began with the opening line, "It was a setup! And if Blacksburg merchants are embarrassed to have gotten caught, they should be."[3] A photo of Krista and Karly sitting behind the mound of cigarette packs they had purchased illegally accompanied this front-page editorial.

Could media attention make a beneficial difference? We answered that question empirically. Back went Krista and Karly into the same twenty stores to buy a pack of cigarettes. The media attention did make a positive difference, but not as much as we had hoped. Karly was now unsuccessful at seven of twenty stores and Krista was turned down at four of the twenty stores. When I shared my disappointment in these results, both daughters took me to task. They told me of being hassled by schoolmates who claimed it was now more difficult for them to purchase cigarettes.

Plus, Krista recounted a particularly uncomfortable incident. She approached the cashier of a large grocery store as a single pack of cigarettes moved along the conveyor belt. Suddenly she saw the front-page photo of her sister and her taped to the cash register with the boldly-printed message, "Do not sell cigarettes to these girls!" Before Krista could make her escape, the cashier recognized her from the newspaper and gave her a public tongue-lashing that was both embarrassing and humiliating.

Later, Krista served as a research assistant (RA) in a variety of additional contexts, from systematically observing the behaviors of alcohol-impaired students in controlled-drinking studies to completing a critical behavioral checklist (CBC) on drivers and employees in various work settings. She continued as an RA in our Virginia Tech (VT) Center for Applied Behavior Systems (CABS) while pursuing a B.S. degree at Radford University (RU), just thirty miles from VT, with a double major in psychology and media studies.

Upon graduating from RU, Krista earned her M.S. and Ph.D. degrees in Human Development at VT with a concentration in family studies and a certificate in gerontology. She taught several undergraduate classes at VT before joining the corporate world as a culture change agent, trainer, and consultant for BBS, PBS, and AC4P. She has traveled the world to teach principles and applications of AC4P to corporate leaders, safety professionals, front-line supervisors, and wage workers. I am humbled and grateful that Krista is following in the footsteps of her dad—*a dream chaser*.

It is most gratifying that we have coauthored this leading-edge teaching/learning manual. Krista's successes and failures at teaching and applying AC4P principles in companies worldwide have provided invaluable connections to real-world practice, thereby making this training guide most applicable.

## The AC4P Movement

In 1991, I coined the term *actively caring* in my editorial for the *Journal of Applied Behavior Analysis*,[4] and since then my SPS partners and I have helped organizations worldwide cultivate an AC4P culture to keep workers injury free. Not only does this approach employ AC4P coaching, but also the humanistic concepts of self-esteem, self-efficacy, belongingness, personal control, and optimism, as addressed later in this manual. Research has shown that people's propensity to perform AC4P behavior increases when these five person-states are enhanced.[5] We consider our AC4P approach to be *humanistic behaviorism*—applying the principles of humanism to make behaviorism more effective.

Following the April 16, 2007, tragedy at VT when a gunman took the lives of thirty-two students and faculty, the concept of AC4P gained a new focus and prominence for our university community. In a time of great uncertainty and reflection, those most affected by the tragedy were not thinking about themselves, but rather acting to help a classmate, friend, or stranger. This collective effort became the ultimate manifestation of the AC4P movement, promoted by the ac4p.org website where AC4P books and wristbands can be purchased.

## The AC4P Wristband

For almost fifteen years prior to the VT tragedy I had distributed green silicone wristbands with the message "Actively Caring for People" in order to promote a culture of employees looking out for the safety and welfare of each other with an interdependent mind-set. After the deadliest school shooting on a U.S.

college/university campus, this wristband took on new meaning. Each wristband was numbered sequentially so the passing of an AC4P wristband could be tracked worldwide over the Internet.

This AC4P process is termed SAPS, for See, Act, Pass, and Share. People who receive an AC4P wristband are asked to pass it on after noticing an AC4P behavior of another person (See). They then reward that AC4P behavior with the wristband (Act), tell the recipient they shouldn't keep the wristband but reward another person's AC4P behavior with the wristband (Pass), and finally (Share) the AC4P story at the ac4p.org website, along with the number on the wristband.

To date, more than 5,000 AC4P stories have been shared at this website, and more than 150,000 AC4P wristbands have been purchased, with proceeds supporting the intervention research and scholarship of CABS. Recently, the College of Science at VT established the Alie Reichling Memorial Scholarship in loving memory of Alie Reichling, who was a dedicated and founding member of the AC4P movement. This special scholarship awards $2,500 each semester to an undergraduate psychology major who has demonstrated extraordinary commitment to the AC4P movement.

Moreover, the AC4P wristband is currently used in police departments nationwide to build positive police-citizen relations throughout communities,[6] and schoolteachers and school resource officers (SROs) are using the AC4P wristband to prevent interpersonal bullying and increase empathy, compassion, and cooperative learning.[7] For AC4P Policing, the wristbands are blue, and police officers share stories of AC4P behavior they observe among citizens at a special website: www.ac4ppolicing.org.

Moreover, I have been distributing green AC4P wristbands at my presentations and workshops for regional and national safety conferences for more than thirty years. A few organizations have reported company-wide efforts to use the AC4P wristbands to successfully increase occurrences of safe behavior and cultivate an AC4P work culture.[8] I hope this teaching/learning manual will inspire the application of AC4P wristbands to reward safety-related AC4P behavior and promote participation in the AC4P movement.

## AC4P Terminology

The reviewers of an initial draft of this manual suggested we include a glossary of the key terms we use throughout the text. After selecting the terms and then beginning to draft definitions, I realized the value of this suggestion. In fact, I wish my prior books that addressed the human dynamics of injury prevention had a glossary of key terms.

The language of psychologists is quite different from that of other professionals (e.g., engineers, physicians, and safety professionals). In fact, behavioral scientists commonly use terminology uncommon in other domains of psychological science. More specifically, my colleagues, students, and I have coined certain AC4P terms and distinct ways of defining some basic psychological concepts. Thus, this teaching/learning manual contains a glossary of eighty-seven key terms used frequently throughout this text, along with definitions from our ABS and AC4P frameworks (see pp. 147–154).

Readers with minimal or no background in psychology will find it useful to read this list of terms and definitions before delving into the content. You can refer to this glossary when coming across particular AC4P jargon in the text, or when writing or discussing the answers to questions designed to activate critical thinking and action planning relevant to a particular lesson.

Even if you have had one or more courses in psychology or experienced a workshop on the human dynamics of safety, you will benefit from reviewing the glossary. Why? We define each term with a behavioral and/or AC4P focus, and quite often this particular focus results in distinct differences from more common usage of the term. In other words, we present operational (or observational) and functional definitions through the lens of an applied behavioral scientist.

You might decide to skip the glossary for now and dive into the content of this teaching/learning manual. Learn how behavior-based safety (BBS) first began, and then evolved to people-based safety (PBS), and currently to actively caring for people (AC4P). Behavioral science has not changed, but extensive experience and evidence-based, objective results have informed best practices to teach and apply the science of improving behavior for sustaining safety, health, and human welfare.

This teaching/learning manual provides the best topical and practical information on applications of behavioral science to achieve an injury-free workplace. You will learn how to cultivate an interdependent AC4P culture of people intervening routinely to keep each other safe.

My daughter and I are thrilled you have our book. Please read the text, discuss answers to the dialogue-provoking questions, and practice the lesson-engaging exercises. Your reward will be fewer injuries on and off the job and active participation in expanding the AC4P movement—spreading interpersonal compassion and behavior-based coaching beyond the workplace to benefit much more than safety. Thank you for actively caring for people.

E. Scott Geller, Ph.D.
Alumni Distinguished Professor and Director
Center for Applied Behavior Systems
213 Williams Hall
Virginia Tech
Blacksburg, VA 24061-0436

# ACKNOWLEDGEMENTS

For forty years, I (Scott Geller) have taught the evidence-based principles and applications of AC4P (or humanistic behaviorism) in workshops and keynote addresses at regional and national safety conferences, and I have documented these in numerous books, manuals, book chapters, magazine columns, and professional research journals. While these presentations might have been effective at teaching the human dynamics of improving safety at some level, two limitations have reduced the large-scale ownership and implementation of the research-based information delivered.

First, many presentations (i.e., in books, book chapters, and journal articles) were too academic and detailed to hold the attention of readers looking for practical application strategies. Who has time to wade through a lengthy book that takes up more space describing the theory and research underlying a process than detailing relevant action plans? And second, when specific behavior-change techniques are presented, they often do not fit a reader's particular needs and/or the context in which they are needed.

These limitations have been expressed in evaluations of my presentations at numerous safety conferences, with statements such as, "I appreciate the theory and principles presented by the professor, but I don't know how to apply his teaching." Or, "I am impressed with the amount of information presented and the numerous possibilities, but I don't know what to do with all of this. How can I get my people engaged in using this information?"

As an AC4P people-based safety trainer, human performance change agent, and leadership consultant throughout the past decade, Krista has heard similar concerns. Numerous times she challenged me with the same basic drawback, "Dad, participants claim to understand the basics of BBS, but they seem to lack the authentic leadership and empowerment to implement a BBS plan. We need to create something people can use in the field and directly apply to their unique safety concerns and issues."

This education/training handbook addresses this legitimate concern in the best way possible. How? By not only explaining the essence of BBS (which has evolved to AC4P), but by providing questions, dialogue, and practice exercises to activate the customization of evidence-based AC4P principles for a specific jobsite. In other words, this manual instructs and inspires readers to develop and apply an effective AC4P plan for their culture, and their distinctive safety concerns.

We want to recognize and thank the many safety leaders who have raised the critical issue of practical application over the years, and thereby incited the preparation of this handbook. These inspirational individuals include the staff of Safety Performance Solutions, who have been teaching the AC4P principles and applications covered in this manual to organizations worldwide since 1994. Their queries and insights from the field not only indicated the need for this manual but also influenced its contents.

Many others have inspired the development of this handbook, verifying the need for a teaching/learning process that enables participants to customize principle-centered AC4P for their culture and providing inspiration and/or guidance. Among these supportive colleagues, a few stand out: Susan Bixler, John Drebinger, Dave Johnson, Nancy Kondas, Charlie Morecraft, and Martin Ralph.

Several special people have substantially influenced Krista's years in the business world, but three have been most inspiring. First, the late Linda Berryman kicked off Krista's safety career and taught her the courage to "stop a job" when safety is compromised. Dave Allen and Bill Redmon worked with Krista throughout her employment at Bechtel as dependable and supportive mentors and colleagues.

We are also grateful for the creative and instructive illustrations by George Wills of Blacksburg, Virginia, and for the word processing and text format support of Brian Doyle and Ashley Underwood.

Last but certainly not least, we are appreciative of the competent support of the publication staff at the American Society of Safety Engineers, as well as the constructive feedback we received from seven anonymous reviewers of an earlier version of this teaching/learning manual. We are particularly grateful for the book project and editorial leadership of Rick Blanchette and the text design and composition of Cathy Lombardi.

Thank you all very much, including our many supportive colleagues and friends whose names are not listed above. The synergy from your advice and inspiration empowers a laudable legacy: AC4P principles and practices readers can use to enrich their lives and contribute to cultivating a work culture where AC4P behavior is the social norm and an injury-free workplace is expected and common.

# Introduction

## BEHAVIOR-BASED SAFETY TO ACTIVELY CARING FOR PEOPLE

The first author used the term behavior-based safety (BBS) in the late 1970s when researching applications of behavioral science to increase the use of vehicle safety belts throughout Blacksburg, Virginia, the home of Virginia Tech (VT). Research students took to the streets with pens and clipboards and obtained empirical records of drivers' use of safety belts before, during, and after the implementation of a community-based intervention. All of the interventions were positive, in contrast to the current ubiquitous disincentive/penalty approach of "Click It or Ticket."

"If-then" incentive/reward contingencies were set up throughout the town. Banks and fast-food restaurants gave customers lottery coupons or bingo numbers at the drive-by windows if they were buckled up. Local merchants donated various prizes for winners of lotteries and "BELTS Bingo." When directing traffic on campus, police officers used pocket-size tape recorders to document the license-plate numbers of drivers who were buckled-up. Later these numbers were translated into lottery coupons for weekly Get Caught Buckled Up drawings. Each week, the VT President, William Lavery, drew ten winners, whose names were published in the local newspapers, with their prizes and the names of the businesses that donated them.

This behavioral-community approach offered positive consequences for all. Banks and fast-food restaurants got more patrons using their drive-by windows; local businesses were recognized in local media for their contributions; and most important, drivers significantly increased their use of safety belts.

These win-win interventions, along with positive media attention, more than doubled the use of safety belts throughout the Blacksburg community and VT campus.[9] Word spread beyond Blacksburg. The U.S. Department of Transportation and the National Highway Traffic Safety Administration (NHTSA) funded several follow-up studies of this positive approach to improving road safety.

In 1982, the Corporate Safety Director for Ford Motor Company requested this application of BBS principles to increase the use of vehicle safety belts among Ford employees at approximately 110 faculties. This led to the development of training materials to teach change agents how to **D**efine a target behavior (in this case, vehicle safety-belt use), **O**bserve and record its frequency of occurrence, **I**ntervene with positive prompts, incentives, and rewards to increase the frequency of safety-belt use, and then to **T**est the impact of the intervention by comparing frequencies of belt use during baseline, intervention, and withdrawal phases.[10] This was called the DO IT process. Safety-belt use among all Ford employees increased from 9% to 54% in 1984, saving at least 20 lives, reducing injuries to more than 800 other-vehicle occupants, and saving an estimated $22 million.[11]

Following the success of this corporate-wide BBS program, the corporate safety director asked Scott to expand the BBS principles and procedures to address the human side of occupational safety within Ford plants. This led to Scott delivering BBS workshops at the facilities of several automobile-manufacturing companies, including Ford, General Motors, and Chrysler. Then the safety professionals of various plants represented at these workshops implemented BBS interventions at their facilities to increase the frequency of safe behavior and decrease occurrences of at-risk (or unsafe) behavior among wage workers.

Every intervention process included some form of behavioral observation and feedback, but the operations varied significantly. Plus, some BBS programs included an incentive/reward intervention whereby various types of rewards (e.g., baseball caps and shirts with a safety logo, lottery coupons, and gift cards) were offered for specific safety-related behaviors (e.g., completing a specific number of observation/feedback sessions, reporting an analysis of a *near miss* or the correction of an environmental hazard, or achieving a certain group percent-safe goal from the CBC).

Unfortunately, the methods and results of these BBS programs were not documented for formal dissemination, but success stories were communicated at safety conferences and contributed to making the BBS approach extremely popular.

## The Popularity of BBS

In the 1990s, BBS became increasingly popular in industrial settings nationwide, and subsequently throughout the world. Our sampling of the research literature identified reports of successful BBS interventions well beyond the U.S. to the UK,[12] Kuwait,[13] China,[14] and Hong Kong.[15]

The general BBS procedures followed by these authors were consistent with the behavioral observation-and-feedback principles detailed later in this manual and included: (1) defining safe and unsafe target behaviors and developing a critical behavioral checklist (CBC) to use during an observation session; (2) applying this checklist when observing and recording relevant safe and unsafe (or at-risk) behaviors of individual workers; (3) following the observation session (which varied from one to fifteen minutes, depending on the number of behaviors observed) with a discussion of the results between the observer and the employee observed. Often a percent-safe score is calculated after the observation session by dividing the number of safe behaviors by the total number of observations and multiplying by 100.[16]

The nature of the peer-to-peer feedback conversations cannot be determined from most of the research publications, but we strongly recommend a nondirective approach where the CBC results are shown to the individual observed for his or her comments.[17] The observer does not give directions, but rather offers behavioral feedback from a fact-finding and AC4P stance.[18] For example, instead of telling a worker certain behaviors observed were unsafe, the observer might ask, "Is there a safer way to do that task?" We provide more details about this nondirective (humanistic) approach later under Lesson 3.

A critical aspect of each post-observation discussion is the identification of contextual and management-system factors that could have influenced the occurrence of at-risk behavior, as well as those factors serving as potential barriers or inhibitors of safe behavior. While observing the employee, the observer notes these potential determinants of safe versus at-risk behavior in the comments column of the CBC. Specifying these factors during the post-observation feedback session facilitates the identification of hazards, conditions, or contingencies that would improve workplace safety if changed.

Most often the results of each observation session are entered into a computer program capable of comparing percent-safe scores across work areas and/or teams. Sometimes the observer's statements in the comments column are entered into a computer program for a subsequent content analysis and identification of common facilitators of at-risk behavior and barriers to safe behavior. The aim is not to find fault with any individual or work team, but to discover behavior and conditions that need follow-up attention and intervention.

## Deviations from the Principles

The popularity of analyzing and summarizing the CBC data with computer-assisted technology is noteworthy. Why? Because it often leads to excessive focus on the data-sheet checkmarks in lieu of constructive peer-to-peer feedback conversations about the observed behaviors and conditions that put people at risk for personal injury.

Geller and Veazie note this deviation from an optimal application of the CBC.[18] They claim the primary purpose of formal observation-and-feedback sessions is not to obtain behavioral data for a computer program, but rather to give employees practice at talking to each other about their safe versus at-risk behavior under certain circumstances.

It's hoped such formal behavioral observation-and-feedback communication will lead to daily informal coaching for workplace safety, eventually resulting in a brother's/sister's keeper AC4P culture in which employers continually look out for the safety of themselves and others with an interdependent mind-set.

The evolution to a culture in which workers "have each other's back" is impossible if supervisors conduct the behavioral observations, and this is often the case. Why? Because it's efficient at getting data for a computer program and does not require costly education and training of line workers.

Too often, even the work supervisors do not receive the necessary education to understand the research-based principles underlying a behavioral observation-and-feedback process. Nor do they receive proper training on how to complete a CBC and deliver behavioral feedback so it's accepted and appreciated by the individual observed.

Most published reports of a BBS process indicate that employee *training* was a component of the intervention, but the nature of the training is rarely specified. Was the format a lecture or an interactive discussion? Was there sufficient role playing and behavioral feedback to justify the label "*training* session"? Was the *education* sufficient to teach the rationale of the BBS process and thereby convince participants of its meaningfulness?

Indeed, it's likely many of the BBS consultants from around the world are not even aware of the applied behavioral science (ABS) foundation of BBS, as reviewed in this manual. As a result, they present incomplete and narrow perspectives of BBS. Too often they offer a quick-fix, step-by-step behavior-change program rather than the ABS philosophy and principles from which a behavior-improvement process can be customized for a particular concern, circumstance, context, and culture.

## The Challenge of Sustaining BBS

The protocol of the BBS interventions reported in the research literature varies dramatically, and we found limited demonstrations or discussions of long-term behavior change. For example, no research report documented a transition from formal to informal behavioral observations and feedback, as advocated by Geller and Veazie.[18] Although we are aware of several companies that have continued a successful BBS process for several years, the long-term effectiveness of BBS is not reported in the research literature. We do have firsthand information of the long-term impact of one BBS process as described below. It is noteworthy, however, that the process evolved to PBS and then to AC4P.

At the 2014 annual safety conference sponsored by Safety Performance Solutions, the six facilitators of a behavioral-safety process for the 1,500 employees of an oil-drilling company in southern California reported annual results of their observation-and-feedback process which was initiated in 1997.[19] Specifically, these individuals showed graphs of man hours, reported injuries, and total recordable injury rates (TRIR) per year for nineteen consecutive years. As depicted in table 1, annual man hours did not vary significantly over this time, but the number of reported injuries and TRIR decreased rather successively and dramatically after a peer-to-peer AC4P coaching process was initiated in 1997.

The AC4P facilitators who reported the impressive long-term outcome data in table 1 indicated the following as key to their remarkable success: (a) teaching all employees the procedures and the rationale of these procedures; (b) maintaining open relationships with the field workers and keeping them in mind at every level of our efforts; (c) promoting interpersonal trust and the theme that

| Year | Number of Man Hours | Number of Injuries | TRIR[a] |
|---|---|---|---|
| 1995 | 2,750,000 | 82 | 6.26 |
| 1996 | 2,900,000 | 80 | 5.47 |
| 1997 | 3,100,000 | 71 | 4.41 |
| 1998 | 3,000,000 | 50 | 3.35 |
| 1999 | 2,500,000 | 53 | 3.86 |
| 2000 | 3,000,000 | 60 | 3.99 |
| 2001 | 2,950,000 | 39 | 2.64 |
| 2002 | 2,500,000 | 28 | 2.22 |
| 2003 | 2,500,000 | 24 | 1.94 |
| 2004 | 2,600,000 | 21 | 1.62 |
| 2005 | 2,750,000 | 30 | 2.21 |
| 2006 | 2,850,000 | 23 | 1.06 |
| 2007 | 2,900,000 | 21 | 1.43 |
| 2008 | 3,200,000 | 19 | 1.16 |
| 2009 | 2,300,000 | 8 | 0.68 |
| 2010 | 2,800,000 | 11 | 0.78 |
| 2011 | 3,200,000 | 13 | 0.79 |
| 2012 | 3,900,000 | 10 | 0.52 |
| 2013 | 4,400,000 | 13 | 0.59 |

[a]TRIR is a measure of the rate of recordable workplace injuries, normalized per 100 workers per year, and is derived by multiplying the number of recordable injuries in a calendar year by 200,000 (100 employees working 2,000 hours per year) and dividing this value by the total man hours actually worked in the year.

**Table 1.** Annual outcome results from a company that has sustained peer-to-peer BBS observation-and-feedback since 1997.

behavioral observation and feedback occurs for the benefit of all employees; and (d) validating employee ownership of the AC4P coaching process by making it visible that workers' peers are involved in every aspect of AC4P safety, from training new employees, upgrading and refining materials and procedures, and analyzing and interpreting the results from the CBCs.

## The Rationale for this Teaching/Learning Manual

So again we come to the purpose of this manual. As indicated in the preface, we have found no published guide to teach/guide employees through the basic principles of BBS, PBS, or AC4P and help them customize the evidence-based lessons for their work culture. Surely, many consultants have workbooks for their education/training, but these are not readily available to the public.

More important, the variety of BBS, PBS, and AC4P workbooks we've examined—including our own—are either too academically complicated or too extensive to be conveniently applicable in the workplace, or they are too narrow and specific. Many describe one particular application of relevant principles—a program promoted by the consultants as *their* best way to apply the concepts. Or, perhaps the program reflects only education or training, but not both.

### Safety *training* or safety *education*?

When we teach a university class it's considered *education*—even *higher education*, but when we teach for industry, the activity is commonly referred to as *training*. Colleges and universities have Centers for Educational Excellence while corporations have training centers. Thus, it seems that colleges and universities *educate*, and industries *train*.

Is there a meaningful difference between education and training, or can we use these terms interchangeably? Obviously, we do use these terms interchangeably, as if they have the same meaning. But if you think about it, these terms are not the same. Each implicates a different teaching style and a different purpose. Some situations require education, others call for training, and some need both education and training. As safety teachers, we need to know when to educate, when to train, and when to do both. Let's explore this distinction further.

First, let's understand the difference between education and training. Actually, you already know the difference. What do you want for your teenager in public school: sex *education* or sex *training*? In contrast, are you satisfied with your teenager receiving only driver *education*, or would you prefer some *training* with that education?

### Misuse of terms

Because people know intuitively the difference between education and training, misusing these terms can have unfortunate psychological effects. Safety training, for example, can come across as a step-by-step procedure or program with no room for individual creativity, ownership, or empowerment. This can lead to the perception of safety as a top-down flavor of the month, here until the next fad comes along. Thus, when people are not educated with regard to the principles or rationale behind a particular safety policy, program, or process, they might participate only minimally. They will perceive the program as a requirement rather than an opportunity to make a difference.

By the same token, safety education without follow-up training will not reap optimal benefits. Learning the theory or principles behind an intervention approach is crucial for customizing intervention procedures for a particular work situation. But after the procedures are developed—optimally with input from an educated work team—training is necessary.

People need to know precisely what to do in order to carry out a set of program procedures. With proper education, these participants can refine or upgrade procedures when appropriate. And with a change in procedures, additional training is obviously needed.

### Different techniques

Teaching methods are different for education than they are for training. When we lecture to large groups of university students or to safety professionals and hourly workers, we use a variety of techniques in an attempt to maintain attention and get participants involved in the learning process. We might use brightly colored overheads or PowerPoint slides, write statements on a blackboard or flipchart to make an extreme statement to elicit contrary reaction, or ask pointed questions and solicit answers from the audience.

Our purpose is to influence the participants' cognitive or thinking processes. From a practical perspective, we want to enhance their ability to improve situations or solve problems related to the subject matter. Such education might increase profound knowledge or critical thinking skills and this could lead to behavior change. If so, we will have *thought a person into behaving differently*. In other words, education targets thought processes directly, and might indirectly influence what people do.

Training, however, targets behavior directly, and might indirectly influence thought processes. This typically calls for more than a lecture format, as described above for education. Although training might start with a specification of the steps needed to accomplish a particular task, more than this is needed to assure the development of certain skills or procedures.

Participants in a training course should practice the desired behavior and receive pertinent feedback to support what's right, and correct what's wrong or could be improved. And if such feedback is given appropriately (as detailed in this manual), behavior might not only be directly improved, but one's thinking or attitude associated with the behavior might be positive. In this case, training would perhaps *act a person into different thinking*.

This manual provides both education and training. Therefore, the explanation of each research-based principle for AC4P intervention is followed by questions or scenarios to facilitate group discussion. Behavioral exercises are given to practice each principle and receive supportive and corrective feedback for continuous improvement. The sharing of opinions and ideas will illustrate the variety of relevant applications from one research-based AC4P principle.

Some of these group discussions will become brainstorming sessions of innovative applications for AC4P intervention. And when some of these possibilities are practiced through interpersonal role playing with feedback, you will have genuine *training* that increases the probability of beneficial applications of AC4P principles to cultivate an injury-free workplace.

## Actively Caring for People (AC4P)

It's imperative to understand the difference between caring and acting. From our conversations with thousands of employees in the corporate world, it's clear that no one wants to see anyone get injured on the job. This is *caring*. Yet, when asked to muster the courage to say something to a peer who is working at risk and could be hurt, the majority admits they do not *act* on their caring by giving their coworkers corrective feedback. Nor do many feel comfortable providing personal appreciation when observing a coworker's safe behavior. Thus, these employees do not *actively* care. They don't go beyond their intrinsic caring to proactive action.

At a power plant in Maryland, thirty-one-year-old Benjamin Heron fell 150 feet to his death in August 2007 after wearing his safety harness but not tying off when he moved on the platform from one tool to another. Two coworkers on the platform noticed he had unhooked and then did not tie back on. But these workers said nothing, and Benjamin fell to his death moments later, slipping between the platform and the first rail.

There are too many stories like this, where chances to help are passed up and the ultimate price is paid. It's not that Benjamin's coworkers didn't care; it's that they didn't *act* on their caring. They said nothing to remind Benjamin to tie off. His wife and three children likely wonder every day why someone didn't go beyond caring to acting by giving Benjamin corrective feedback. Of course, they had their excuses (e.g. "We're only contractors and aren't supposed to tell full-time employees what to do"), just as we all rationalize our lack of proactive AC4P behavior. It's likely, however, these men do not experience many days without thinking, *If only I had . . . .*

**From BBS to AC4P**

Many companies have implemented an AC4P process for safety called BBS[20] or people-based safety (PBS).[21] Its purpose is to help people develop the confidence and skills to take care of each other every day (i.e. to be brother's/sister's keepers) before injuries occur—to be continuously proactive for safety. As indicated above, employees use a critical behavior checklist (CBC) to pinpoint safety-related risks to look for when systematically observing each other.[22] Then they observe their coworkers and check "at-risk" or "safe" on the CBC, based on the behavioral definitions they previously determined in a group meeting with their work team.

After completing the CBC, the observer provides supportive feedback to the observed worker for the safe behaviors recorded and corrective feedback for any at-risk behavior noted. This behavioral feedback is not given to direct behavior change, but to empower personal acceptance and self-accountability. A nondirective or humanistic approach to feedback delivery is most likely to be appreciated and considered.[23]

The CBCs are given to a data processor who tallies the information and distributes a summary to all employees. At-risk behaviors occurring at the highest frequencies are identified, specifying targets for further improvement. This information is shared with all employees so together they can learn where more mindful focus is needed for injury prevention. The purpose: to cultivate an AC4P culture in which 100% of the time, every day, any worker seeing another worker at risk (including breaking a safety rule) speaks up to prevent the possibility of injury.

The real value of using a CBC is to help people develop in confidence and competency, so they provide feedback every time they see someone at risk of being injured. The formal observation-and-feedback process helps participants become more skilled and comfortable at providing and receiving interpersonal feedback regarding safe and at-risk behaviors. Such formal observations are actually completed as practice to nurture people's competence and courage to provide behavioral feedback any time they see someone in potential danger of an injury.[18]

## From formal to informal observations

Consider a winning athletic team. Rigorous long-term practice with behavioral feedback occurs before the first game. With BBS, the practice is formally using a CBC to improve behavior. As with athletics, the practice is important but the purpose of practice is preparation for the game.

The essential intent of the formal BBS process is to enable employees to *play their best game*. Do not assume your number of formal observations completed will lead to fewer on-the-job injuries. Only when the observations occur voluntarily, beyond the formal practice sessions, will the connection between the behavioral observation-and-feedback process and injury rate most likely occur. The ultimate leading indicator: How often do workers give *informal* feedback to sincerely commend ongoing safe behavior or warn a coworker of a hazard or an at-risk behavior they observe?

Such interpersonal discussions are routine in an AC4P culture, but we have found few work cultures with this level of actively caring. But routine AC4P behavior for safety is essential to sustain an injury-free workplace. This teaching/learning guidebook will help you reach this ideal AC4P culture. Let's consider how five person-states influence a person's propensity to actively care for safety.

## Preparing people for AC4P behavior

Consistent proactive AC4P behavior for safety can be difficult. It's usually inconvenient and it's not always well-received. The science of psychology explains why we might choose to act on behalf of the well-being of another person, or why we might avoid an opportunity to actively care.

Research findings from psychological science reveal that five person-states influence one's propensity to perform AC4P behavior.[24] These five evidence-based person-states—called "states" rather than "traits" because they change as a function of various situational and interpersonal factors—are depicted in figure 1. Connections between these person-states and AC4P behavior are as follows:

- Self-esteem: The better I feel about myself, the more likely will I serve others.

- Personal control: The more choice and personal control I perceive in a certain situation, the more likely will I intervene on behalf of another person's welfare.
- Self-efficacy: The more perceived competence I have to intervene effectively, the more likely will I actively care.
- Optimism: The more optimistic I am that my AC4P behavior will be helpful, the more likely will I intervene.
- Belongingness: The greater the personal connection I feel with the person whom I perceive could benefit from my AC4P behavior, the more likely will I intervene.

The critical point about these person-states is that they change as a function of situational factors, especially interpersonal conversations.[25] If a supervisor builds relationship connections by appreciating or asking about an employee's weekend, this leader enhances one or more of these person-states, and increases the likelihood that individual will perform AC4P behavior.

Many managers know the power of *management by walking around*.[26] But it's not the walking around that's important; it's the one-to-one conversations with employees. An AC4P manager converses to boost one or more of the five person-states. If all they do is walk around and tell people what to do and point out shortcomings, they can deflate rather than inflate one or more of these person-states.

**Figure 1.** Five person-states that influence propensity to perform AC4P behavior

Consider this: What if one of those managers stopped and helped, or even picked up some trash or assisted with the housekeeping in an area? This kind of perception lets others know that housekeeping, picking up trash, and helping to stay organized on the site is not below a manager. In fact, the manager is actively caring for that team by lending a hand. This creates a constructive relationship, one that provides a positive perception for managers and leaders.

Leaders with an AC4P mind-set understand the importance of building constructive relationships with employees. They are sensitive to the five person-states and realize how the nature of their conversations with others can impact these states in positive or negative directions, thereby helping or hindering the cultivation of an AC4P culture. AC4P is not just making noise around people or getting in their space. It's a matter of showing genuine concern for them—their five person-states and the culture in which they work. This level of leadership generates self-motivation and fuels an AC4P workplace.

We have given many workshops on how to cultivate an AC4P safety culture, and we have received numerous practical suggestions from participants on how they enhance each of these person-states in others. Some of these suggestions are reviewed briefly below, selected from more comprehensive and detailed presentations.[24]

Please note the special value of soliciting strategies from your own work teams by asking three questions per each person-state: (1) What happens around here that lowers the person-state? (2) What happens around here that enhances the particular person-state? And (3) What can we do in our workplace to boost this person-state in others?

### Self-Esteem

Participants have suggested a number of ways to build self-esteem, including: (1) provide opportunities for personal learning and peer mentoring; (2) increase one-to-one recognition for desirable behaviors and individual accomplishments; and (3) solicit and follow up on people's safety suggestions.

It's important to give more positive (or supportive) than negative (or corrective) feedback. It's essential to focus on the act, not the actor, when offering corrective feedback. An error only reflects behavior that can be corrected, not some deeper character flaw. Don't come off as a judge of character, implying a mistake suggests some subjective personal attribute such as carelessness, apathy, bad attitude, or poor motivation.

Be a patient, active, and empathic listener. Allow people to offer reasons for their error or poor judgment. Don't be tempted to argue about these explanations—it's just a way for them to protect their self-esteem, and it's generally a healthy response. Remember, you already made your point by showing the error and suggesting ways to avoid the mistake in the future. Leave it at that.

It might help to explore feelings if a person doesn't react constructively to corrective feedback. You might ask, "How do you feel about this?" Listen with empathy to assess whether self-esteem has taken a hit. You'll learn whether some additional communication is needed to focus squarely on what is external and objective, rather than internal and subjective.

Leave the situation on a positive note. Take into consideration that a positive interaction will enable the person to retain the corrective feedback you have offered and will engender good feelings about the entire one-on-one communication.

### Self-Efficacy

Self-efficacy is more situation-specific than self-esteem, so it fluctuates more readily. Job-specific feedback should be directed only at the behavior needed to perform a particular task successfully. It should not veer off in the nebulous direction of general self-worth. Keep in mind that repeated negative feedback can have a cumulative effect, chipping away at an individual's perception of self-worth. Then it takes only one remark, perhaps one you would think is innocuous and insignificant, to "break the camel's back" and activate an overreaction.

Our communication may not be received as intended. We can do our best to come across positively and constructively, but due to factors beyond our control, the communication might be misperceived. Indeed, one's person-state can dramatically bias the impact of interpersonal feedback. Note that self-efficacy reflects the perception of competence, and when people receive sincere supportive feedback about their exemplary safety-related behavior, their perceived competence is enhanced, fueling their self-motivation.[27]

### Personal Control

Employees at our seminars on AC4P have listed a number of ways to increase perceptions of personal control, including: (1) set short-term goals and track progress toward long-term accomplishment; (2) offer frequent rewarding and correcting feedback for process activities rather than only for outcomes; (3) provide opportunities to set personal goals, teach others, and chart small wins[28]; (4) teach employees basic behavior-change intervention strategies (especially feedback and recognition procedures); (5) provide time and resources for people to develop, implement, and evaluate intervention programs; (6) show employees how to graph daily records of baseline, intervention, and follow-up data; and (7) post response-feedback graphs of group performance.

The perception of personal control is analogous to the perception of personal choice or autonomy. When people believe they are in control of a situation or challenge, they generally feel a sense of personal choice. "I choose to take charge of the mission which is within my domain of influence." Appreciate the similarity between these person-states. People are more self-motivated when they perceive choice or a sense of autonomy.[29]

### Optimism

Optimism flows from thinking positively, avoiding negative thoughts, and expecting the best to happen. Anything that increases self-efficacy should increase optimism. Also, when personal control is strengthened, people perceive more influence over their consequences. This gives them more reason to expect the best.

Note how the person-states of self-efficacy, personal control, and optimism are clearly intertwined. A change in one will likely influence the other two. Note also how these person-states relate to perceptions of choice and competence, which are determinants of self-motivation.

### Belonging

Here are common proposals for creating and sustaining an atmosphere of belonging among employees, as suggested by participants at our discussion groups: (1) decrease the frequency of top-down directives and quick-fix programs; (2) increase team-building discussions, group goal-setting and feedback, and group celebrations for both process and outcome achievements; and (3) facilitate self-managed or self-directed work teams.

Feelings of empowerment and belonging can be enhanced when groups are given control over important matters, such as developing a behavior-improvement observation-and-feedback process or a particular AC4P initiative. When resources, opportunities, and talents enable team members to assert that "we can make a difference," feelings of belonging occur naturally. This leads to synergy, with the group achieving more than could be possible from participants working independently.

## Personal Impact

You can impact your own five person-states. Take self-esteem, for example. When you help another person, you feel better about yourself. You feel more valuable because you helped another person do better and/or feel better. Plus, helping others can boost your other person-states, from optimism to belongingness.

What if you have an opportunity to perform an act of kindness for someone, but you don't perform the obvious AC4P behavior? Might you feel guilty? Could any of your person-states be lowered, such as your sense of personal control, self-efficacy, optimism, and/or your sense of belonging? Understanding these possibilities indicates you're on your way to becoming an AC4P leader.

Bottom line: When we act to serve others, the victory is not just theirs, it's ours as well—we create a win-win relationship. Their win is obvious; they get our help. Our win is a boost to one or more of the five AC4P person-states, which in turn increases the likelihood we will continue to serve others.

However, when we choose not to actively care for another person's safety, we not only lose an opportunity to enhance our own levels of self-esteem, belongingness, optimism, self-efficacy and/or personal control, we likely lower one or more of these positive dispositions for ourselves. Hence, we have a potential self-reinforcing AC4P cycle. When you act on your caring, you enhance your own five beneficial person-states. And the higher these person-states, the greater the probability you will perform another AC4P behavior, and boost those critical person-states again.

When you study, discuss, and role-play implications of the seven fundamental lessons explicated in this teaching/learning handbook, you will note numerous direct connections between the lessons, the five person-states, and ways to enhance these states in yourself and in others. All for one purpose: to increase the frequency and improve the quality of AC4P behavior on behalf of people's safety, health, and well-being.

# Lesson 1

## EMPLOY MORE POSITIVE CONSEQUENCES

First, realize that behavior is motivated by consequences. As Dale Carnegie, author of the seminal book *How to Win Friends and Influence People*, said in 1936, "Every act you have ever performed since the day you were born was performed because you wanted something."[30] We do what we do because of the consequences we expect to gain, or escape, or avoid by doing it.

Question: Would you rather be influenced by positive consequences or negative consequences? Your personal experience and common sense provide the answer to this question, and it's verified by over sixty years of behavioral research.

What is your reaction to the illustration below?

Will the athlete run? Of course he will. People react unhesitatingly to avoid the threat of an impending negative consequence. But how long will the athlete run when the coach is not around to hold him accountable? What will be his attitude about running? Will he become self-motivated to run without the need for an extrinsic consequence to keep him going?

What about the wage worker who is reprimanded harshly and publicly by a supervisor for not wearing certain personal protective equipment (PPE)? Yes, this employee will immediately don the required PPE. But how will the employee feel? Embarrassed with lowered self-esteem? Will the employee feel self-motivated to use this PPE when the supervisor is not around? Might this employee actually show off nonuse of the PPE in order to assert personal freedom to other wage workers?

We are not suggesting the nonuse of PPE should be overlooked. Corrective feedback is certainly called for when safety rules are not followed, but a top-down negative approach can do more harm than good. Later, under Lesson 3, we address the administration of corrective feedback in ways that promote mutual respect, personal buy-in, and even self-motivation.

We cover the important topic of self-motivation later in this manual, including ways to enhance the self-motivation of yourself and others. For now, just consider the potential unpleasant effects of using the threat of negative consequences to activate avoidance behavior and the administration of a negative consequence following the occurrence of an undesired behavior.

The most efficient and effective way to improve both behavior and attitude at the same time is to follow desirable behavior with positive consequences. In fact, what happens to one's attitude when undesirable behavior is followed by a negative consequence? We all know the answer to this rhetorical question, right? But if we realize the attitudinal and behavioral advantages of using positive rather than negative consequences to influence behavior, we must ask: *Why are negative consequences used more often than positive consequences to improve behavior at work, in school, at home, and throughout communities?*

**Please record your answers to this question here and then participate in a group discussion about answers, if possible.**

_______________________________________________

_______________________________________________

## A Matter of Mind-set

Note how the answers to the prior discussion question reflect this typical finding: The use of negative consequences is convenient and the behavioral impact is usually immediate. But discuss answers to this question. *Which behavior-change technique, the application of positive or negative consequences, will have the longest-term beneficial impact? Why?*

Often, the same situation can be viewed as: (a) control by penalizing unwanted behavior, or (b) control by rewarding desired behavior. Some students in our university classes, for example, are motivated to avoid failure (i.e., a poor grade), while other students are motivated to achieve success (e.g., a good grade or increased knowledge).

Which students feel more empowered and in control of their class grade? Which ones have a better attitude toward the class? Of course, you know the answers. Reflect on your own feelings or attitude in similar situations where you perceived your behavior was influenced by either positive or negative consequences. Figure 2 depicts four distinct achievement-related person-states that have been researched by behavioral scientists to explain differences in people's attitude and motivation when working to achieve success versus avoid failure.

It's most desirable to be a *success seeker*. These are the optimists who respond to setbacks in a positive and adaptive manner. They are self-confident and willing to accept challenges, as opposed to evading demanding tasks to avoid

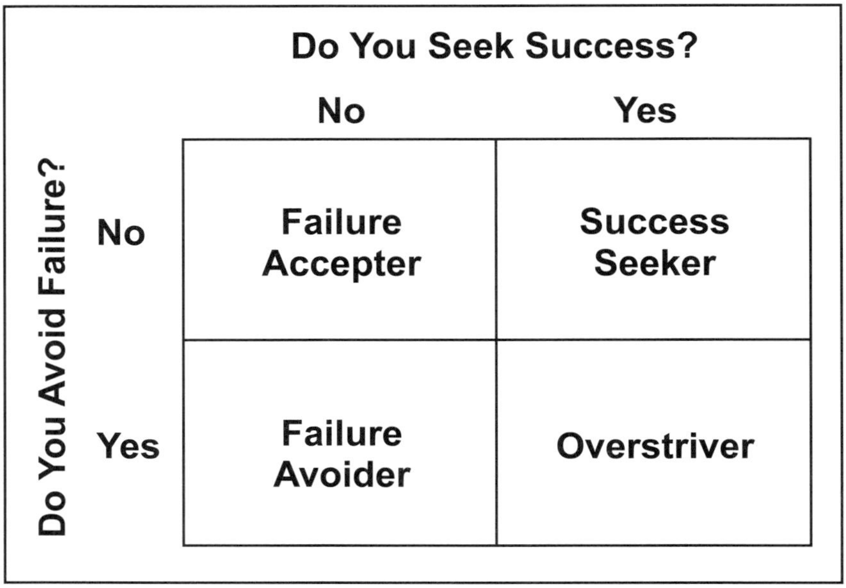


**Figure 2.** Four achievement-related person-states defined by achieving success or avoiding failure

failure. They wake up each day to an *opportunity* clock rather than an *alarm* clock. It's a mind-set or attitude toward life you can encourage in yourself and others with situational manipulations and communication, both interpersonal and intrapersonal (i.e., self-talk).

*Failure avoiders* have a low expectancy for success and a high fear of failure. They do whatever it takes to protect themselves from appearing incompetent. They often set low expectations and use defensive pessimism to shield themselves from experiencing failure. These individuals are motivated but are not happy campers. They are the students who say, "I've *got* to go to class; it's a *requirement*," rather than, "I *get* to go to class; it's an *opportunity*."

**What kinds of conversations do you have with others and to yourself that influence a success-seeking over a failure-avoiding mind-set, and vice versa?**

Applying soon, certain, and positive consequences is the most effective way to improve behavior and attitude at the same time. Using this life lesson daily is both critical and challenging. Why? Because we seem to live in a "click it or ticket" culture that relies on negative consequences to manage behavior from the workplace and classroom to our homes, and during our travel in between. Indeed, the usual government approach to behavior management is to pass a law and enforce it.

Compliance safety and health officers from OSHA (Occupational Safety and Health Administration) tour industrial sites to look for safety violations. Financial penalties are imposed for violations of OSHA rules. Plus, the safety professionals of many organizations serve as safety cops, looking for individuals and work teams who violate company safety policy. Violations are punished, often with a formal reprimand in the form of a letter. And after three violations for a safety rule, the employee is sent home for a day without pay.

**Given the disadvantages of using negative consequences or punishment to manage behavior, do you think the safety-cop approach to injury prevention has become less prominent? Why or why not?**

**What kinds of actions can be taken to eliminate the safety-cop perception during a behavioral observation?**

Bottom line: Sometimes behavioral observers are perceived as snitches. This kind of perception can be detrimental to AC4P coaching and can stifle sincere engagement. So, it is imperative to always provide a positive observation without a lot of directives and negative statements.

It is also crucial to say "thank you" after every observation and feedback session. Some companies provide a thank-you sticker or safety coin to give to the employee after the AC4P coaching session. A soon, certain, and positive token of appreciation certainly increases the acceptance of the AC4P coaching team. Of course, each team member has been educated on the invaluable benefits of behavioral observation and feedback, and they have practiced how to verbalize these to others.

## Discussion Questions

1. **Describe a situation when you delivered, received, or observed the application of a negative consequence as a reaction to observing a safety violation. What was the immediate behavioral impact?**

2. **Do you believe this punishment tactic was justifiable in this case? Why or why not?**

## When is Punishment Appropriate?

Before delving out a negative consequence for an undesired behavior or outcome, answer the following seven questions. In most cases, you'll find another type of corrective action is more appropriate.

**1. Was a specific rule or regulation violated?**

If you answer no, punishment is obviously unfair. Does this mean you need to write more rules or document more regulations? We don't think so. You can't write a rule for every possible unsafe behavior. And since human errors are unintentional, rules won't decrease them. We need to allow for the possibility that noncompliance with a rule or regulation can be unintended. This leads to the next question.

**2. Was the behavior intentional?**

All human error is unintentional. We mean well but have cognitive failures or brain cramps. Psychologists call these errors slips or lapses, and they are typically due to limitations of attention, memory, or information processing.[31]

Have you ever walked into a room and forgot why you were there? Locked yourself out of your car with the keys still inside? Left your house only to return to get something you forgot? Looked for your cell phone while talking on it? In the workplace, research has shown these types of errors actually increase with experience on the job. Skilled people put their actions on automatic pilot, and perhaps add other behaviors to the situation. How many of us juggle a cellular phone or read a text message while driving? You can see how an error can easily occur in such situations.

This *unconscious* incompetence needs to be corrected, but certainly not through punishment. Unfortunately, too many of us are rewarded when we take the risk and read our text message while driving. We discuss later how it often takes an emotional communication to become aware of our risky behavior and make the conscious choice to substitute safe for unsafe behavior.

We also have *conscious* incompetence. Sometimes poor judgment is used to intentionally take a risk. This conscious behavior can be classified as either a

mistake or a calculated risk. Have you ever miscalculated a parking space and scraped an adjacent vehicle? Pressed the brake pedal too quickly on a slippery road or pumped the brakes in an antilock system? Parking and braking are frequent and intentional driving behaviors, but under particular circumstances they are mistakes.

Now suppose you don't buckle your safety belt. You know this behavior is unsafe, but you decide to take a calculated risk. In this case, unlike a mistake, you're aware that your behavior is inappropriate. The deliberate or willful aspect of calculated risks might seem to warrant punishment, but punishment won't convince people their judgment was defective. And it's this recognition that's needed to change conscious incompetence to conscious competence.

This cartoon above illustrates the all-too-common at-risk behavior reflecting conscious incompetence. At least 99% of us realize that talking on a cell phone or texting while driving is an at-risk behavior, but many of us still take this risk. It is a temptation many of us engage in deliberately. So how do we change this reinforced behavior? The same issue occurs when workers take risky shortcuts. The risky behavior is reinforced, and is usually not followed by a negative consequence—an injury.

Would it help to add a contrived negative consequence like a penalty for the intentional at-risk behavior? Perhaps the undesired behavior will change temporarily, but not over the long term. And the negative impact on attitude can be detrimental to self-accountability for safety.

We offer alternatives to punishment later in this guidebook. For now, just consider that people do perform at-risk behavior with intention, but such behavior is usually reinforced by natural positive consequences. If these positive consequences cannot be eliminated or reduced, consider ways to make the alternative safe behavior more reinforcing before adding a negative consequence for these undesired behaviors.

### 3. Was a rule knowingly violated?

Researchers have proposed that the more knowledge or skill we have at doing something, the less likely we are to demonstrate poor judgment.[32] So many of the driving mistakes discussed above occur often from inexperience or poor training. On the other hand, the tendency to take a calculated risk increases with experience on the job. This is human nature, and it won't be changed with punishment.

Some errors occur because the rule or proper safe behavior was not known. And it's possible for an experienced worker to forget or inadvertently overlook a rule. Training and behavior-based observation and feedback can reduce these types of errors, but punishment certainly won't help in this case.

For example, if a worker forgets his or her gloves or leaves his or her hard hat in the lunchroom, it's important to recognize that this behavior is usually not done on purpose, but is just a simple brain cramp. Overreacting with punishment is not helpful. Increasing the behavioral reminders around your jobsite is helpful. For example, hang signs in areas where hard hats and gloves might be removed—the lunchroom, the bathrooms, in and around offices. These kinds of signs can serve as reminders and activate the safe behavior.

### 4. Were other employees endangered?

Failing to lock-out a power source during repair work is the behavior most often targeted for punishment. We've heard many safety professionals say, "We only punish employees when a particular behavior risks a severe injury or fatality, or places other individuals in danger." Some safety pros are quick to add that these risks must involve willful intention and prior knowledge.

Now if an employee willingly and knowingly avoids a lock-out procedure to put him and others at risk, then the severest punishment is relevant. Actually, this person should be fired immediately. But this personal rationale for performing such risky behavior is very rare.

Many dangerous behaviors are mistakes resulting from poor judgment, not an unconscious or conscious desire to circumvent policies and hurt someone. And when a calculated risk is taken, it's not performed with the expectation that someone will get hurt. Often specific characteristics of the work environment or culture enable or even encourage a calculated risk.

**5. What factors support the behavior?**

This is the most important question of all. People don't make errors or take calculated risks in a vacuum. Poor judgment occurs for a reason. And it's important to learn an employee's rationale for taking a risk. This enables a truly useful corrective-action discussion. Did the employee lack knowledge or skills? Was a demanding supervisor or peer pressure involved? Did equipment design invite error with poorly labeled controls? Was the safe way inconvenient, uncomfortable, or cumbersome?

Let's consider the organizational culture: Is safety taken seriously only after an injury? Is safety performance judged only in terms of injuries reported per month, instead of the number of preventive activities?

These are only some of the questions that need to be asked. Punishment will only make answers—and a practical corrective action plan—harder to come by. Punishment creates a negative attitude toward corrective action and stifles the type of open conversation needed to develop an effective plan for improving behavior.

**6. How frequent is the behavior?**

How many at-risk behaviors typically occur before an injury occurs? H. W. Heinrich estimated three hundred close calls per one major injury,[33] and Frank Bird observed this ratio to be six hundred to one.[34] Whereas the hypothesized frequencies are different, both Heinrich and Bird presumed numerous at-risk behaviors occur before even a close call is experienced, let alone an injury.

So, what good is it to punish one of many risky behaviors? If the behavior is an error, punishment will only stifle reporting and potential remedies. If the probability of getting caught while taking a calculated risk is low—and it's minuscule if you wait until an injury occurs—any threat of being punished will have little behavioral impact. Remember, punishment reduces self-motivated engagement in safety improvement efforts more than it reduces the occurrence of at-risk behavior.

### 7. How often have others avoided punishment?

Punishing an employee for a behavior that others have performed without receiving similar punishment is one sure way to lose credibility and turn a person against your safety efforts.

As an AC4P team member, you should never take it upon yourself to punish someone. If you are a witness to a behavior you feel puts others at risk, have a supervisor or manager continue the discussion with the employee. If an AC4P process leader has that negative discussion, you might create the safety-cop mentality and lose the trust you have developed for the AC4P team and its approach to injury prevention.

In summary, if the threat of personal injury is not sufficient to motivate consistent safe behavior, it doesn't help to add one more threat (punishment) to the situation. We need open and frank discussions with the people working at risk in order to analyze and change management practices, equipment, or organizational systems that contribute to much of the at-risk behavior we see in the workplace. This is only possible when the threat of punishment is removed. Figure 3 depicts a classification of undesirable behavior, and indicates when punishment might be advisable.

| | Culture Encouraged: Yes | Culture Encouraged: No |
|---|---|---|
| **Intentional: Yes** | **Calculated Risk**<br>***No Punishment*** | **Calculated Risk**<br>***Punishment May Be Warranted*** |
| **Intentional: No** | **Preventable Slip, Lapse, or Mistake**<br>***No Punishment*** | **Unpreventable Slip, Lapse, or Mistake**<br>***No Punishment*** |

**Figure 3.** When punishment for at-risk behavior may be warranted

## Discussion Questions

1. To what degree does your work culture use the threat of punishment (or a negative consequence) to manage safety? In other words, how is behavior managed by the threat of punishment in your work culture? What unsafe behaviors are apt to be punished if they occur at your workplace?

2. To what extent are unsafe behaviors, minor injuries, and the factors contributing to these topics of conversation among members of your work team? In other words, how often do workers talk about their close calls, minor injuries, and the relevant contributing factors?

3. What characteristics of your work culture contribute to stifling critical conversations about at-risk behavior and minor injuries? What are the factors that contribute to these barriers to an injury-free work culture?

**4. What could be done to increase the occurrence of frank and open discussions about at-risk behaviors and conditions in your workplace and about ways to decrease these precursors to an injury?**

**5. We suggest the consequence following an observed at-risk behavior should be a *corrective feedback conversation* with the employee. Who should have this conversation and what should be covered in such a conversation? Later in this teaching/learning guide we discuss ways to give corrective feedback so the conversation is most beneficial for everyone involved. For now, use your current common sense, but your answers may change after reading Lesson 3 on how to deliver behavioral feedback.**

# Lesson 2

## BENEFIT FROM OBSERVATIONAL LEARNING

If you want to be better at what you do, watch someone who performs that behavior better than you do. Of course, we all realize the power of observational learning. Indeed, a large body of psychological research indicates this type of learning is involved to some degree in almost everything we do.[35] Whenever you do something a particular way because you saw someone else do it that way, or because someone showed you how to do it that way, or because characters on television or in a video game did it that way, your actions are the result of observational learning.

Whenever you attribute peer pressure as the cause of someone taking up an unhealthy habit (like smoking cigarettes or drinking excessive amounts of alcohol) or practicing an at-risk behavior (like driving at excessive speeds, taking a shortcut, or adjusting equipment without locking out the energy source), you are referring to observational learning.

When you remind someone to set an example for others, you are alluding to the critical influence of observational learning. Think of that manager or supervisor walking the site and picking up trash, helping a team organize its work space, or assisting with some housekeeping in another area. That kind of example-setting behavior activates observational learning and benefits everyone who sees it. It also contributes to a positive perception of that manager.

As children, we learned numerous behavioral patterns by watching our parents, teachers, and peers. For instance, when we saw our siblings or schoolmates receive rewards such as special attention for certain behaviors, we were

more likely to copy that behavior. This process is termed vicarious, or indirect, reinforcement. At the same time, when we observed others getting punished for performing certain behaviors, we learned to avoid those behaviors. This is referred to as vicarious punishment.

As adults, we teach others by example. As the illustration above shows, our children learn new behavior patterns, including verbal behaviors, by watching us and listening to us. In this way, they learn what is expected of them in various situations.

The prominent role of parental modeling in socializing children is illustrated by this humorous but true story of a father tucking his six-year-old daughter into bed:

> When a father came into his daughter's bedroom, she requested, "Daddy, would you tuck me in like you do Mommy every night."
>
> He said, "Sure, honey," and pulled the covers up and underneath his daughter's chin.
>
> As he left the room, his daughter called after him saying, "Wait, Daddy. Would you give me a good-night kiss like you do Mommy every night?"
>
> Dad said, "Sure, honey," and he kissed his daughter on her cheek.

As he was leaving the room again, his daughter called after him with one more request, "Daddy . . . Daddy, wait. Would you whisper in my ear like you do Mommy every night?"

"Sure honey," he replied, and he leaned down and went "Bzzz, bzzz, bzzz" in his daughter's ear.

Then she said, "Not tonight, Daddy, I have a terrible headache."

Our actions influence others to a greater extent than we realize. Without us being aware of our influence, children learn by watching us at home, and our coworkers are influenced by our practices at work. Not only do the occurrences of safe acts encourage similar behavior by observers, but verbal behavior can also be influential. If a supervisor is observed commending a worker for her safe behavior or reprimanding an employee for an at-risk behavior, observers may increase their performance of similar safe behaviors (through vicarious reinforcement) or decrease the frequency of similar at-risk behavior (through vicarious punishment).

Indeed, to make safe behavior the norm rather than the exception, we must always set a safe example both in our own work practices and in the verbal consequences we offer coworkers following their safe and at-risk behaviors. In other words, you never know when you are being observed. So, walking

the safety talk and performing the safe options benefit the observers through observational learning. The three illustrations below provide a memorable pictorial regarding the influence of example-setting on observational learning and behavior.

## Discussion Questions

1. **Explain how on-the-job training involves observational learning, including vicarious reinforcement and vicarious punishment.**

2. **How much of your current job did you learn by watching the behavior of others? Please explain.**

3. **Have you ever learned an at-risk shortcut for your job by watching another employee? Please explain. If so, did you take the shortcut yourself or did you consciously decide to play it safe? Why or why not?**

4. A common internal excuse for taking a risk is, "It won't happen to me." And the experience of most individuals supports this rationale. People take many risks, such as texting while driving, and they don't get hurt (i.e., they get to their destination without a vehicle crash.). How could you use this observational learning lesson to alter this excuse for performing an at-risk behavior? In other words, given the powerful influence of observational learning, the internal dialogue ("It won't happen to me") should be changed to what?

5. Please explain how particular safe behavior on your job could influence the behavior of other employees on your team.

# Lesson 3

## IMPROVE WITH BEHAVIORAL FEEDFORWARD AND FEEDBACK

"Practice makes perfect" is not true; practice makes permanence. Only with behavior-focused feedback can practice improve behavior. Sometimes feedback is a natural consequence, as when line workers observe the finished product—the outcome of their labor. But even when we observe the natural results of our behavior, behavioral feedback from an observer (e.g., a safety professional or AC4P coach) is necessary for proper behavioral adjustment and improvement.

People want to be competent at tasks they believe are meaningful or worthwhile, and doing something safely is always the most competent way. Managing a workplace to keep employees safe and injury-free is certainly worthwhile work. So how can workers become more safe and competent on the job?

The answer: relevant, behavior-focused feedback. But who should deliver such improvement feedback? Employees—members of a work team—need to give each other supportive and corrective feedback. The letters of COACH say it all: Care, Observe, Analyze, Communicate, Help (see figure 4).

The coaching process starts with caring. "Know that I *care*, and you'll care what I know. I care so much, I'm willing to *observe* you and notice the occurrences of desirable (or safe) and undesirable (or at-risk) behavior."

The observer also notes environmental factors that may be influencing the observed behavior, from environmental conditions to behavioral consequences. This is the *analysis* phase of coaching.

***Care***

- *show that you care*
- *set caring examples*

***Observe***

- *define target behaviors*
- *record behavioral occurrences*

***Analyze***

- *identify existing contingencies*
- *identify potential contingencies*

***Communicate***

- *listen actively*
- *speak persuasively*

***Help***

- *recognize continuous improvement*
- *teach and encourage the process*

**Figure 4.** The five components of AC4P coaching

Next, we have interpersonal *communication*, the delivery of information gained from the prior *observation* and *analysis* steps. Most people want to improve, but many people resist giving and receiving the kind of communication critical for beneficial behavior change.

Some people perceive feedback that implies personal change as an indictment of their current work style, job skills, or diligence. This reaction is most likely to occur when someone is asked to change dramatically and when current procedures have been followed for years.

To overcome resistance to behavior-based feedback, effective behavior-improvement coaches steer clear of disruptive and dramatic communication, rather emphasizing incremental fine-tuning or successive approximations. They also facilitate beneficial change to both behavior and attitude by accentuating the positives—the occurrences of safety-related behavior.

If the interpersonal communication goes well, the last component of COACH—*Help*—is accomplished. The behavioral feedback was accepted and will be used to improve the pinpointed behavior. Note how the four letters of HELP—Humor, Esteem, Listen, and Praise—reflect strategies to increase the probability that a coach's advice, directions, or feedback will be appreciated, accepted, and followed.

## AC4P Coaching for Safety

AC4P coaching is essentially an interpersonal process of one-on-one observation and feedback. One person, the coach, systematically observes the behaviors of another person and then provides constructive feedback related to these observations. Safety coaches support the safe behaviors they see and offer corrective feedback regarding any at-risk behavior they observe.

Let's consider ten guidelines for initiating an effective AC4P coaching process throughout an organization. There is solid empirical evidence that this approach to injury prevention works to maintain pro-safety behavior and prevent workplace injuries.[36] These ten guidelines or rules for establishing and maintaining an effective interpersonal AC4P coaching process for injury prevention were not derived overnight, nor were they gleaned from research articles or textbooks.

Rather, they were developed from more than two decades of studying more than five hundred actual industrial applications. Hence, these guidelines can be considered lessons learned from real-world experience helping organizations initiate and sustain an effective behavioral observation-and-feedback process. They reflect the state of the art in behavior-based AC4P coaching.

### 1. Teach procedures with principles

Earlier, we distinguished between education and training. Education explains *why*, and training shows *how*. Motivation to learn what to do—the procedures—can come from understanding the underlying rationale—the principles. Before people are trained on how to conduct behavioral observation and feedback, they should be educated on the research-based foundation of AC4P safety.[37] Then the participants can appreciate the procedures that put the principles in practice.

When participants learn and accept the principles behind a safety initiative, they can help define and refine tools and techniques applicable for their work groups. Such participant involvement in designing process steps facilitates empowerment and ownership, the next guideline.

### 2. Empower employees to own the process

Genuine empowerment is not given. It's enabled and then released from people when they feel ownership.[38] But ownership does not come easy, and it doesn't happen overnight. And ownership is not the same as compliance.

Employees often comply with rules, regulations, and operating procedures without ownership. In this case, they perform a certain way because someone is holding them accountable. Such behavior is *other*-directed.[39] Ownership, on the other hand, implies internal control, *self*-accountability and *self*-directed behavior. We're talking about self-motivated behavior, and we expand on both empowerment and self-motivation later under Lesson 5.

People who get involved in designing, implementing, evaluating, and refining a process acquire a special degree of ownership. Their contributing behaviors are self-directed. These behaviors occur consistently because participants hold themselves responsible, and not because someone else is holding them accountable.[40]

How can this level of ownership be reached? The remaining guidelines are essential to making this happen. For example, the next implementation principle specifies that participants need to exercise some personal choice throughout an AC4P coaching process, from designing and implementing initial procedures to evaluating and refining the protocol for continuous improvement. Ownership implies personal choice. And people get more involved in procedures that are influenced to some degree by their personal input.

### 3. Provide opportunities for choice

Choice, involvement, and ownership go hand in hand. Each supports the other two. More of one influences more of the others. Furthermore, choice is motivating. Research has shown that even insignificant choice benefits commitment and relevant behavior. For example, people have shown improved performance when they select aspects of a task that are actually irrelevant to effective completion of the task.[41]

So how much choice is optimal? Is it possible to allow too much choice in an AC4P process? One systematic evaluation of twenty successful behavior-based safety (BBS) programs indicated that too much choice can be detrimental. More specifically, DePasquale and Geller found that BBS programs labeled *completely voluntary* were generally not as successful as BBS programs introduced with the explicit *expectation* that everyone will get involved to some degree.[42]

Those programs that incorporated an accountability system to track involvement obtained the most participation and success. However, all the most successful BBS coaching programs included some element of choice throughout process development, implementation, and continuous improvement.[43]

Maintaining an effective balance between external accountability and personal choice is reflected in this general management principle: Provide structure and direction, but accompany your advice with opportunities to select among alternative action plans. In other words, management should provide structure, instruction, and support for occupational safety, while providing opportunities for employees to develop and choose among procedural options.

Management should always maintain a consistent level of support and involvement for a proactive AC4P process, as reflected in the next guideline for implementing and sustaining a behavioral observation and feedback process.

### 4. Facilitate supportive involvement from management

Many consulting firms have marketed BBS as employee driven and management independent. As a result, some organizations have launched BBS tools and methods without active support from management. After arranging for BBS training, the supervisory staff steps back and lets an employee steering committee direct the implementation of a behavioral observation and feedback process.[44] While this enables substantial perception of choice among line workers, employee involvement is typically not optimal.

Whether considering AC4P behavioral coaching or another occupational safety process, a hands-off policy is not optimal. Let's face reality. People give priority to those aspects of their job that get attention from supervisors and managers. In other words, people do what they believe they need to do in order to please those who control their monetary compensation for successful job performance.

Yes, self-directed behavior is best, but often behavior must start as *other*-directed. In most organizational cultures, people need to be held accountable for carrying out the observation-and-feedback process before they can appreciate the natural supportive consequences of AC4P coaching. Plus, supervisors and safety pros can do a number of other things to encourage and support AC4P coaching, including:

- Allocate time to discuss process activities and results at group meetings
- Contribute to group discussions of AC4P coaching procedures and results
- Help schedule and coordinate opportunities for AC4P coaching activities, such as observation-and-feedback sessions
- Request systematic observation and feedback for certain tasks
- Use the observation data to identify environmental hazards and barriers to safe behavior
- Help remove hazards and barriers identified in the behavioral-observation-and-feedback process
- Request periodic briefings on data from the AC4P coaching process, such as the amount of participation, percent-safe behavior, number of coaching sessions performed, percentage of safety suggestions implemented that had been activated by AC4P coaching, and the results of special AC4P intervention efforts
- Recognize individuals and teams for their notable AC4P coaching participation
- Organize and support group celebrations of special safety achievements

**5. Ensure the process is nonpunitive**

The last two bullets of Guideline 4 above refer to the use of recognition and group celebrations to support AC4P coaching activities and accomplishments.

This guideline avoids the use of negative or punitive consequences, as we discussed under Lesson 1. Indeed, researchers have demonstrated specific disadvantages of traditional enforcement procedures.[45] Here the critical point is that connecting negative consequences to any aspect of an employee-driven (and management-supported) AC4P activity can kill the entire process. Punishment stifles feelings of interpersonal trust, empowerment, ownership, and commitment.

The data from a behavioral-observation-and-feedback process reveal at-risk behaviors and environmental hazards needing attention. They can also demonstrate less than optimal participation in a critical safety-related procedure. Such negative results, or specification of improvement needs, can provoke an enforcement mind-set and suggest a need for punitive consequences. It's essential to retreat from this traditional approach to safety management.

We are not recommending the elimination of all punishment or discipline applications, even though most of these are not corrective and probably do more harm than good. Managers who use negative consequences to motivate compliance do so at their own risk. But an enforcement policy must be administered independently of all peer-to-peer coaching activities.

The workforce must believe the data from their AC4P coaching process cannot be held against them. Finding low participation or remarkable at-risk behavior is not cause for a negative consequence; rather, it pinpoints opportunities for improvement. Open and frank conversations about areas of concern are much more likely than a negative consequence to increase mindful commitment to change and to activate peer support for specific improvement targets.

### 6. Ensure the coach is nondirective

At first, peer-to-peer observation and feedback can feel awkward for both the observer and the observee. In fact, the behavioral coaching process can come across as confrontational, with one person (the observer) assigned to audit another person's work practices and then offer corrective advice for eliminating any at-risk behavior observed. Such a perception of behavioral coaching hinders interpersonal trust and stifles involvement, ownership, and empowerment.

From the start, it's critical to emphasize that the AC4P observer (unlike a typical athletic coach) is not responsible for corrective action. The observer merely completes a critical behavior checklist (CBC) and shows the observee the results. In fact, the CBC had been developed through interactive group discussion among line workers representative of the relevant workforce.

The two workers might discuss environmental or system factors that discourage safe behavior and encourage at-risk behavior. And, they might consider ways to remove obstacles to safe behavior. The AC4P coach offers relevant positive words of approval to recognize certain safe behavior but gives neither disapproval statements nor direction related to any at-risk behavior observed.

In some situations, the observer might notice some useful safety technique. If this is the case, the observer should spend time praising the observee for his or her safe example and thank him or her for the opportunity to learn another safe way to perform the task. But even if no supportive behavioral feedback is warranted, the observer should sincerely thank the observee for his or her time and assistance with the observation-and-feedback process.

With regard to at-risk behavior, the coach is nondirective.[46] In other words, the observer provides specific behavior-based feedback for the observee to consider. There is no peer pressure to change. The only accountability is self-accountability. Any adjustment in behavior is self-directed, provoked by the results of a nonintrusive and anticipated application of the CBC.

### 7. Progress from announced to unannounced observations

Consider the word *anticipated* in the prior sentence. Taken literally, it means the recipient of an observation-and-feedback session knows it's coming and can prepare for a good showing. Consequently, the behavioral observations are not random, and the results are not really representative of a worker's daily routine. The CBC data are biased toward the positive. The percent-safe score is higher than reality warrants.

The next guideline builds on this point about unrealistic (or invalid) behavioral data. Here we need to consider a justification for announcing the behavioral observations. If making employees aware of the observations leads to overly positive results, why announce them? One way to answer this question is to consider the alternative.

Imagine workers sneaking around and completing CBCs unbeknownst to those being observed. This approach would be viewed by many as a "gotcha program" and would undermine interpersonal trust, involvement, and ownership. The lower percent-safe scores from this approach might be more accurate, but at the expense of the attitudes and person-states needed to achieve the interpersonal cooperation and learning requisite for achieving an injury-free workplace.

Another technique would be to approach the work group from a distance, taking note of any at-risk behaviors you are witnessing and then making it a point to discuss those behaviors during the interaction phase of the observation. For example, in order to maintain a positive discussion, you could say, "I noticed from a distance, as I approached. . . ." This would allow you to discuss the at-risk behaviors you saw without seeming like you were hiding in the bushes.

Even when they know they are being observed, workers still perform certain at-risk behaviors. These are actually the work practices that benefit most from behavioral feedback. The observation process holds people accountable to perform their jobs as safely as they know how. When they learn ways to be safer under these circumstances, workers truly add new behavioral patterns to their knowledge base. This is invaluable behavior-based learning.

While this guideline reflects the need to start AC4P coaching with announced observations, a transition to unannounced observations is clearly optimal.[18] Specifically, the organizations most successful at AC4P coaching progress from announced to unannounced behavioral observations. This shift should only occur when workers realize the process is truly for their own benefit. The guidelines presented here help make this happen.

Some clients of Safety Performance Solutions (SPS) have developed creative ways to facilitate the transition from announced to unannounced observations. For example, one organization incorporated individual choice (Guideline 3) by distributing hard-hat stickers workers could display to indicate their willingness to be observed at any time without being asked. Thus, the workers at this site placed a special sticker on their hard hats if they were willing to be the recipient of an unannounced behavioral coaching session. Eventually, all employees at this facility voluntarily placed this special sticker on their hard hats.

At another facility, employees put their name in a raffle jar whenever they were willing to be observed anytime on a particular day. The observers selected their coaching assignments each day by randomly drawing names from this pool. Eventually, the daily drawings included every worker. People gave permission to be observed when their experience with AC4P coaching convinced them this was not the traditional, top-down enforcement approach to occupational safety, but was an interdependent learning process that enabled workers to actively care for the safety and health of their team members.

The NORPAC paper mill in Longview, Washington, developed an ingenious incentive program that has not only increased personal choice and participation, but has added a fun and constructive diversion to the standard work routine. Specifically, about 10 percent of the mill workers volunteer to be *mystery observees* during intermittent promotion periods. They receive a coupon redeemable for a meal for two at a local restaurant, which they give to the next person who observes and coaches them for safety. Then this employee becomes a mystery observee, anticipating an opportunity to reward another coworker for completing a one-to-one behavioral observation-and-feedback session on him or her.

Each week the employees are asked to complete a CBC on a coworker (with permission) and then communicate the results in a positive one-to-one feedback session. The employees know about the mystery observees but do not know who they are. The process gets people talking about the AC4P coaching process in positive terms, and it rewards the most difficult and most important aspect of the intervention process: interpersonal feedback.

**8. Focus on the interaction, not just the numbers**

The objective data obtained from a comprehensive observation-and-feedback process are a useful component of the process. Computer software helps organize and summarize the results from behavioral checklists and pinpoints targets for intervention. Computer programs can compare various workgroups on various dimensions of an AC4P coaching process and track the results of consecutive days, weeks, or months of behavioral observations. Thus, work teams can benchmark objectively with others and can assess successive attempts to improve quantity and quality of AC4P coaching participation as well as the percentage of safe behavior.

This data-analysis feature of AC4P coaching is critical for its remarkable success. Behavioral data enable objective pinpointing of targets for improvement, as well as continuous evaluation of corrective action procedures.[47] Such data provide objective evidence of accomplishment and thereby justify recognition and celebration. Thus, the data available from AC4P auditing procedures are invaluable, but it's crucial to look beyond the numbers.

It's important to remind people that the numbers are not statistically reliable. They are a collection of subjective information derived from peer-to-peer interaction. The observers are not trained researchers applying the CBC in a verifiable and reliable manner. Rather, these are AC4P volunteers who are applying an observation-and-feedback process to actively care for themselves and their coworkers.

Thus, it's vital to treat the data as a collection of subjective opinions and perceptions from dedicated AC4P team members. Record the data, read the data, understand the data, recognize that the data do tell a story, but realize that story is based on perceptions and does not necessarily reflect reality.

It's easy to become over-analytical with the results of behavioral observations. But the benefits of AC4P coaching extend far beyond the evaluation of CBCs; it's all about the peer-to-peer interaction. As indicated above, most records of behavioral observations are biased and unreliable. They are typically obtained under unnatural conditions, as when the observations are announced beforehand. In addition, there is a tendency to overlook at-risk behavior, especially when an interpersonal feedback session is expected to follow an observation session.

Thus, the data from behavioral-observation-and-feedback sessions provide useful comparative information, across sessions within the same work group and between different work teams, but the absolute values of these numbers should not be taken too seriously. And above all, consider that the process of interpersonal observation and feedback is more powerful than the numbers with regard to achieving an AC4P work culture and an injury-free workplace.

The communication components of AC4P coaching demonstrate the value of peer support, develop interpersonal trust, and help to cultivate the kind of learning-oriented organization that brings out the best in people. The process teaches workers they can be *unconsciously incompetent* and that they need feedback from others to improve, also known as a cold-eyes review. This leads to an interdependent mind-set, a realization that the success of an organization is dependent upon systems of people contributing diverse talents and relying on each other synergistically to make the whole greater than the sum of its parts.

### 9. Continuously evaluate and refine the process

We've mentioned evaluation and refinement throughout this presentation. No process targeting human behavior is carved in stone. Behavior is dynamic, continually adjusting to changing demands, expectations, and ergonomics. Consequently, CBCs need to be revised periodically, along with adjustments to the procedures used to conduct behavioral observations and deliver feedback. Making relevant adjustments to the CBCs increases the likelihood the observers will stay motivated and not become mindless from following a repetitive routine.

With experience, AC4P coaches become more adept at noticing the subtle features of safe versus at-risk work practices, beyond the obvious, such as the use of personal protective equipment. This continual increase in coaching expertise needs to be reflected in revised CBCs. In addition, techniques to support AC4P principles and procedures (such as incentives, accountability techniques, and group meetings) need to be responsive to changes in the workplace, including behaviors, attitudes, management systems, and the environmental context in which work is performed.

Bottom line: It's critical to continually assess the behavioral and attitudinal impact of ongoing AC4P coaching procedures and to make refinements accordingly. An ever-changing process keeps the team, the checklist, the safety practices, and the coaching from becoming stale. An evaluation of people's opinions and attitudes about an AC4P coaching process requires interpersonal conversations with both participants and nonparticipants. These should occur in both group and individual one-to-one sessions.

It's vital to ask people how they feel about the AC4P process: the checklist, the behavioral observations, the coaching, and the identification of factors influencing safe and at-risk behaviors. Regularly soliciting such feedback will continuously improve the beneficial impact of the AC4P process.

Perception surveys can enable a broad site-wide assessment of employees' feelings about an organization's safety culture in general and about an AC4P process in particular. However, a perception survey alone provides no specific direction for procedural refinement. Interviews, focus groups, and team discussions should follow the survey.

It's critical for the employees to feel ownership of the AC4P process. They need to believe they have contributed to the overall state of the process to stay involved and interested. Although this continuous evaluation and refinement takes extra time and effort, the added benefits far outweigh the response costs.

While gaining specific recommendations for improvement, opportunities are provided for employee engagement, choice, and ownership. The value of these dynamic and interdependent qualities of an AC4P coaching process was discussed above in Guidelines 2 and 3.

**10. Make the process part of a larger effort**

Over the years, BBS has attracted many critics. Some of the negative reactions are based on fundamental misunderstandings of the principles and methods of BBS. More specifically, some authors have portrayed BBS as any attempt to influence behavior, regardless of the intervention approach (e.g., policies, training, incentives, and enforcement procedures) or of the intervention target (e.g., attitudes, behavior, or cognitions).[48] Others typify the whole of BBS as only one kind of intervention—behavioral observation and feedback.[49]

While this presentation does focus on behavior-based coaching as a key intervention approach, AC4P principles can be applied to many other domains of occupational safety, including ergonomics, training, recognition and celebration, incident analysis, human-error prevention, hazard identification, and corrective action, to name a few. In each of these cases, AC4P reflects a particular approach toward handling the human dynamics of the initiative or process.

Therefore, observation and feedback is not the entirety of AC4P but rather is an interpersonal AC4P coaching process for improving safety-related behavior that applies certain research-supported principles and tools derived from applied behavioral science (ABS). By reading and discussing the seven life lessons in this teaching/learning handbook, you will learn how to apply ABS and AC4P principles to improving behavior related to human welfare in many domains beyond safety.

It's important to view behavior-based observation and feedback as one of many systematic ways to prevent personal injury in the workplace. Yes, this intervention approach was developed by behavioral scientists, and it incorporates basic principles and procedures from ABS. However, it is neither BBS nor AC4P. Rather, AC4P represents an overall approach toward dealing with the behavioral dynamics of injury prevention.[50]

Just like the guidelines presented here are relevant for the development, application, and evaluation of more safety initiatives than an observation and feedback process, the philosophy and technology of AC4P are applicable to more occupational safety efforts than an observation-and-feedback process, and useful for much more than occupational safety.

## The Misuse of Feedback

Imagine you receive an email from a supervisor requesting that you come to his office at the end of the day to receive some feedback. How would you feel? Would you anticipate a feedback session with unpleasant verbal behavior and emotion? How relevant is the illustration below?

Two common characteristics of feedback influence people's desire to avoid feedback and justify the negative attitude or emotion. First, negative or corrective feedback is typically given more often than positive or supportive feedback is. Many supervisors, teachers, coaches, and parents use reprimands more often than praise, with the apparent belief that we learn more from our mistakes than from our successes. Empirical research and even common sense indicate this is untrue.

The second reason feedback has a negative connotation is that people often correct others without focusing entirely on their behavior. In other words, the feedback delivery suggests the problem or error observed reflects more than behavior: The worker is *careless*; the student is *ignorant*; the athlete is *lazy*; the perpetrator is inherently *evil*; or the child is *sloppy*. That kind of feedback delivery can do more harm than good, and substantial research demonstrates dramatic disadvantages of labeling people beyond their behavior and effort, even when that label is positive.[51]

It's critical to focus on the behavior being performed and not on a subjective label you might connect to what you see. This is easier said than done, but it's essential to separate the behavior from the person when giving and receiving feedback. Corrective feedback is not an indictment of one's personality or an indicator of a character flaw. Feedback must *not* be related to an individual's attitude, motivation, professional competence, or family history. Feedback is only about the observed behavior.

Yes, responding well to supportive or corrective feedback can lead to improvement in attitude, motivation, competence, and even a person-state. But the purpose of feedback is only to pinpoint desired and/or undesired behavior. When this is realized by those who give and receive feedback, the beneficial outcome of behavioral AC4P coaching is maximized. There's room for improvement in most everything we do, and only by receiving and accepting behavior-based feedback can we do better.

## It's in the Delivery

We hope it's clear that giving interpersonal feedback at the right time can certainly increase its beneficial impact on behavior. Actually, timing is one of four basic guidelines we recommend you consider when planning feedback strategies. These rules of feedback delivery can be readily remembered with the acronym SOAR: Specific, On time, Appropriate, and Real. So, by following these four basic rules of feedback delivery, you can *soar* to success with interpersonal behavioral feedback.

### Specific

As indicated above, AC4P feedback needs to focus on specific behavior. As a consequence (for motivation), feedback specifies what behavior to stop and what behavior to keep performing. But as an activator (for direction), it's not feed*back* but rather feed*forward*. Feedforward reminds an individual to perform a particular task in a certain way.

Figure 5 illustrates the activator-behavior-consequence (ABC) model of AC4P. Note the distinction between feedforward and feedback. Both of these behavior-improvement techniques need to be understood and accepted by the relevant person(s).

People can obviously misperceive, misunderstand, or deny AC4P feedforward or feedback. Thus, a *person* is depicted between the Activator (feedforward)

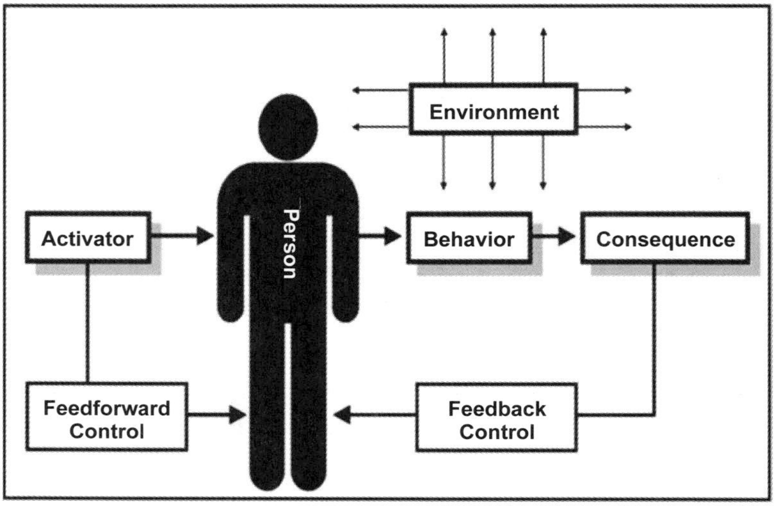


**Figure 5.** The activator-behavior-consequence (ABC) model of AC4P

and Behavior, and between the Consequence (feedback) and Behavior. The principles covered here suggest ways to ensure the acceptance and appreciation of behavioral direction (feedforward) and behavioral support or correction (feedback).

Ambiguous and subjective language about internal person-states is not useful and can be counterproductive. For example, evaluative statements such as, "It seems you're careless, lazy, unenthusiastic, unaware, disorganized, or out-of-touch" will only add resentment and lessen the acceptability of the behavioral message.

It's important to note that when you give positive statements, watch for and eliminate the use of "but." Rather than giving pure praise or appreciation, we often feel obligated to add a negative statement to balance the communication. "I like the way you clarified concepts with real-world and relevant examples, *but* next time try to facilitate more involvement from your audience." Such mixed messages can weaken your impact.

"It's great to see you are wearing the correct cut-resistant gloves and have long sleeves while grinding, *but* I notice you removed the guard on your grinding tool. That can cause serious injury and it would be better to leave the guard in place." Some people will hear only the negative and miss the positive. Others will discount both messages, figuring the net gain of one positive and one negative is no communication.

So, it's often best to make your specific supportive feedback short and sweet. Don't combine both positive and corrective feedback in one exchange or overload a person with several behaviors to continue or change. It's much better to give people brief and specific feedback messages over weeks or months than to give people fewer but longer feedback sessions with mixed and potentially confusing motives and directives.

**On time**

As discussed above, AC4P feedback to increase or decrease the occurrence of behavior should follow the target behavior as soon as possible. An excessive delay between the behavior and the supportive or corrective consequence can be counterproductive, as reflected in the illustration below.

On the other hand, when the purpose of behavioral feedback is to shape the quality of a response, it usually makes most sense to give such directive feedback as an activator (i.e., feedforward) that precedes the next opportunity to perform the target behavior.

Receiving feedback about an error consequently can be perceived as punishing and frustrating if an opportunity to correct the observed error is not available in the near future. When the person eventually receives an opportunity to correct the behavior, the advice might be forgotten. When you provide advice to improve as close as possible to the next opportunity for the behavior to recur (i.e., as feedforward), you increase its directive influence and reduce the potential of a negative attitude resulting from catching a person making a mistake.

Suppose, for example, you observe a coworker walking hurriedly down a flight of stairs without using the handrail. Realizing the relatively high frequency of injuries caused by slips, trips, and falls, you see an opportunity to actively care for this person's safety. Should you provide corrective feedback? Well, it's obvious this individual is in a hurry and would not appreciate a conversation about handrail use at this time. Plus, you might feel awkward providing corrective feedback in this context, with no credible opportunity to buffer the negative with relevant positive words. The solution: Offer behavioral feedforward at another time.

In this case, *on time* means you wait for an opportunity to provide this employee with a feedforward reminder to use the handrail before your coworker uses a flight of stairs. Yes, it may take some time before you can seize the feedforward moment. You might see an opportunity for this pre-behavior reminder when you or the target for your intervention is walking with a group of coworkers. In this instance, you would simply say, "Let's remember to use the handrail."

**Appropriate**

Specific and well-timed feedforward and feedback need to be appropriate for the needs, abilities, and expectations of the person receiving direction, support, or correction. Simply put, the AC4P feedforward and feedback should fit the situation. This means feedforward and feedback should be expressed in language the performer can understand and appreciate, and it should be customized for the performer's skill to perform at the particular task.

Keep the conversation simple and straightforward. Remember KISS when providing feedforward or feedback, "Keep It Simple, Sally." The more complex or detailed the verbal exchange, the less likely the direction or support will be understood and appreciated.

When people are learning a task, directive feedforward and motivational feedback do need to be detailed and perhaps accompanied with a behavioral demonstration. In such learning situations (i.e., observational learning), it's important to match the advice with the performer's achievement level. Don't expect too much and give more advice than the individual can grasp in one feedforward exchange. Give people a chance to fully comprehend the idea before moving on.

At times, risky behavior is performed by experienced line workers who know how to do their job well but they may have developed at-risk habits or are just taking a shortcut for efficiency. It could be insulting and demeaning to give these individuals detailed instructions about the appropriate way to complete their job. In these situations, it's best to give feedforward as a reminder to take the extra time for effectiveness over efficiency. Preceding your reminder with the phrase "as you know" could increase acceptance of the reminder.

Consequently, it's important for the AC4P coach to size up the situation and make specific and timely behavioral feedforward or feedback fit the occasion. This is not easy and requires up-to-date awareness of the performer's knowledge and skill regarding a certain task. It also requires specific knowledge regarding the optimal ways to perform the task in each situation. This is a prime reason why the most effective AC4P feedforward and feedback usually occur between employees on the same work team.

### Real

Another reason interpersonal AC4P coaching is most effective when occurring between team members: It's perceived as most genuine and caring. Feedforward and feedback will be ineffective if the verbal behavior is viewed as a way of exerting top-down control or demonstrating superior knowledge, competence, or motivation. The only reason for giving behavioral feedforward or feedback is to improve and maintain the safe competence of team members.

Although well-intentioned, the discipline policies in some organizations make it difficult for some people to view corrective feedback as caring and supporting. The gotcha mind-set associated with rule enforcement can interfere with a safety professional's sincere attempt to correct at-risk behavior. Remember, you are focusing on the behavior, not the rule. Thus, corrective AC4P feedback is most likely to be perceived as genuine or *real* when it occurs between work peers on the same team. These individuals know most about the situation and the person and thus have sufficient information and opportunity to make the feedforward and feedback specific, on time, appropriate, and real.

## In Summary

We have reviewed key guidelines for delivering effective AC4P feedforward and feedback. In this regard the word SOAR is a useful teaching/learning tool because each letter of this acronym reflects a key word that implicates a rule for delivering feedforward and feedback effectively: Specific, On time, Appropriate, and Real.

These four words indicate the special value of team members giving each other on-the-job feedforward and feedback. Why? Because team members have the most complete knowledge of the feedforward/feedback recipient and the situation. They are best able to make their behavioral feedforward and feedback: (a) *specific* in terms of behavior to initiate, continue, or stop; (b) *appropriate* for the knowledge, abilities, and experiences of the performer; and (c) reflect *real* concern for the individual's competence and job effectiveness.

Members of a work team are also most likely to be *on time* with feedforward or feedback, whether the AC4P communication serves as an activator, motivator, or both. When using feedback to support safe behavior or decrease at-risk behavior, *on time* means the feedback should be delivered as a consequence that follows as soon after the observed behavior as possible. People need immediate feedback if they are going to recognize the at-risk behavior and the change they need to implement in order to set a safe example.

In contrast, when using feedforward to encourage behavior (perhaps as a correction for at-risk behavior observed earlier), you should consider this communication a reminder or an activator. That is, you should wait for a situation that calls for the particular behavior and then offer a specific, appropriate, and genuine reminder.

The life lesson: Give much more supportive (positive) feedback than corrective (negative) feedback; and use feedforward more often than feedback to correct at-risk behavior.

## Discussion Questions

It would be extremely useful to discuss these four guidelines for giving feedforward and feedback in situations and under circumstances relevant for the participants. The format for connecting the principles with realistic applications can vary markedly. The instructor could facilitate commentary from all participants. Or the participants could be divided into small groups, and after the questions and issues posed here are discussed, a representative from each group could report their perspectives to the entire group of workshop participants.

Whatever your protocol for encouraging interactive discussion of AC4P feedforward and feedback, we suggest consideration of the following questions.

In some teaching/learning situations it might be optimal or most convenient for participants to write personal reactions to these questions before a discussion with one or more other participants.

1. **What specific work situations could benefit from AC4P feedforward and feedback between line workers? Is such communication more important under some circumstances than others? Please explain.**

2. **What factors hold employees back from giving each other behavior-based feedforward or feedback regarding safety-related behavior?**

3. **How might the obstacles to giving each other safety-related feedforward or feedback be overcome?**

4. Please discuss personal experiences in which you *received* behavior-based feedforward and feedback from a colleague regarding your safe or at-risk behavior. How were these communication exchanges appropriate or inappropriate with regard to the SOAR guidelines?

5. Please discuss personal experiences in which you *delivered* behavior-focused feedforward and feedback to individuals regarding their safe or at-risk behavior. How were these communications appropriate or inappropriate with regard to the SOAR guidelines?

6. Please discuss situations when you find it particularly difficult or awkward to give feedforward and feedback to someone with regard to safety-related behavior.

**7. How could your typical feedforward and feedback communications with coworkers be improved with application of the SOAR guidelines?**

______________________________________________

______________________________________________

______________________________________________

______________________________________________

______________________________________________

**8. What holds you back from having more effective AC4P feedforward and feedback conversations with coworkers about their at-risk behavior?**

______________________________________________

______________________________________________

______________________________________________

______________________________________________

______________________________________________

Finally, it would be very useful to practice and document occasions of giving feedforward and feedback to colleagues throughout your day. You should note what went well and not so well in your delivery of personal *feedback* for behavioral support or correction. Reflect on these comments as *feedforward* to improve your next opportunity to improve another person's safety-related behavior.

We hope you will use the chart on the following page to systematically document your interpersonal feedforward and feedback experiences. Discussing these experiences with others will be instructive for all as vicarious observational learning (Lesson 2).

## Documentation of Feedforward and Feedback Experiences

| Feedforward or Feedback | Target Behavior | Situation or Context | Positive Outcome | Room for Improvement |
|---|---|---|---|---|
| | | | | |
| | | | | |
| | | | | |
| | | | | |
| | | | | |
| | | | | |
| | | | | |
| | | | | |
| | | | | |
| | | | | |

# Lesson 4

## USE MORE SUPPORTIVE THAN CORRECTIVE FEEDBACK

"We can't learn unless we make mistakes." How many times have you heard this? While this might make us feel better about the errors of our ways and provide an excuse for focusing more on other people's failures than on their successes, nothing could be further from the truth. Behavioral scientists have shown convincingly that success—not failure—produces the most effective learning.[52]

Edward Lee Thorndike, for example, studied intelligence at the start of the last century by putting chickens, cats, dogs, fish, monkeys, and humans in situations that called for problem-solving behavior. Then he systematically observed how these organisms learned. He coined the term *law of effect* to refer to the fact that learning depends upon behavioral consequences.[53] And markedly more learning occurs following positive consequences (success) than negative consequences (failure).

Does an error have to occur in order to solve a problem? You can reflect on your own experiences to answer this question. A pleasant consequence gives you direction and motivation to continue the behavior. You know what you did to receive the reward and are thus motivated to earn another. In contrast, a negative consequence following a mistake only tells you what not to do. It provides no specific direction for problem solving. An overemphasis on a mistake can be frustrating and discouraging and can actually demotivate you to continue the learning process, right?

Errors are not necessary for learning to occur. In fact, when training results in no errors—made possible with certain presentation techniques—learning occurs most smoothly and is most enjoyable.[52] Errors disrupt the teaching/learning process and can lead to a negative attitude, especially if negative social consequences accentuate the mistake. Even subtle reactions to an error—a disappointed face or verbal tone—can increase feelings of helplessness or despair and can turn a person off to the entire learning process.

## Deliver Effective Supportive Feedback

The most powerful positive consequence to support a learning process is supportive feedback. But some attempts to be positive and supportive are ineffective, as the illustration below shows. Five basic guidelines are given next for giving quality supportive feedback.

### Be timely

For supportive feedback to provide optimal direction and support, it needs to be associated directly with the desired behavior. When people know what they did to earn the appreciation, they might be motivated to continue that behavior.

If it's necessary to delay the supportive feedback, the conversation should relive the activity deserving recognition. Talk specifically about the behavior that warrants special acknowledgement. Don't hesitate to ask the recipient to recall aspects of the situation and the commendable behavior. This enables direction and motivation to continue the desired behavior.

### Make it personal

Supportive feedback is most meaningful when it's perceived as personal. This can be done by providing an example of your own learning, your own experiences, your own understanding—all to show you can relate to the topic at hand.

Verbal support should not be generic, fit for any situation, such as, "Nice job!" Rather, it needs to be customized to fit a particular individual and circumstance. This happens naturally when the supportive feedback is linked to designated behavior. When you recognize someone, you're expressing personal thanks. This shows people you truly care, that you are genuine in your conversation. By making it personal, you can identify with the circumstance.

Sometimes it's tempting to say "*we* appreciate" rather than "*I* appreciate," referring to departmental gratitude rather than *personal appreciation*. Speaking for an organization or company can come across as impersonal and insincere

Of course, it's appropriate to reflect value to the organization when giving supportive feedback, but the focus should be personal: "I saw what you did to support the safe behavior of that worker, and I really appreciate it. *Your example demonstrates the kind of leadership we need around here to nurture an AC4P safety culture.*" This second statement illustrates the next guideline for giving quality supportive feedback.

### Take it to a higher level

Supportive AC4P feedback is most memorable and inspirational when it reflects a higher-order quality. Adding a universal attitude such as leadership, integrity, trustworthiness, or AC4P to your recognition statement makes the feedback more meaningful and thus rewarding. It's important to state the specific behavior first and then make an obvious linkage between the behavior and the positive characteristic it reflects.

Later we discuss in more detail the use of an AC4P wristband to show appreciation for safety-related behavior. In these cases, we offer the wristband as a symbol of AC4P leadership, worn to show membership in an elite group of individuals dedicated to cultivating an AC4P safety culture. Thus, the AC4P wristband is given to not only reward AC4P behavior, but also to signify participation in a movement to activate and reward AC4P behavior from your workplace to your home and throughout the community at large.

The conversation that follows the wristband is critical. For the wristband to carry the level of appreciation needed to affect behavior, the person passing out the wristband must deliver an effective one-on-one conversation. Show genuine compassion when delivering the AC4P message and wristband, take your time, and show the other person that this wristband is a symbol of an AC4P movement throughout the jobsite, the entire organization, and beyond.

This connection brings the interpersonal supportive feedback to a higher level, enabling positive impact on the recipient's self-esteem, competence, and sense of interdependence and belongingness. Earlier we explained how these person-states enhance the propensity to perform more AC4P behavior. Later we discuss the connection between these person-states and self-motivation.

**Deliver it privately**

Because supportive feedback is personal and indicative of higher-order attributes, it needs to be delivered privately and one-on-one. This requires a certain degree of courage from those not comfortable in private, one-on-one conversations, especially with people they don't know well. But consider this: Such verbal support is special and only relevant to one person, so it will mean more and seem more genuine if it's given personally from one individual to another.

It seems conventional to recognize individuals in front of a group. This approach is typified in athletic contests and reflected by the pop-psychology slogan "Praise publicly and reprimand privately." Many supervisors take the lead from this common-sense statement and give workers recognition in group settings.

Indeed, isn't it maximally rewarding to be held up as an exemplar in front of one's peers? Not necessarily, because many people feel embarrassed when singled out in front of a group. Part of this embarrassment could be due to fear of subsequent harassment by peers. Some peers might call the recognized individual an apple-polisher or brownnoser, or accuse him or her of "sucking up to the supervisor."

When Scott was in third grade, his teacher recognized him in front of the class for doing an excellent job on his homework. As depicted in the illustration on the next page, Scott was very embarrassed. Then after school, a gang of boys beat him up on the playground. Unfortunately, the teacher never received feedback about the negative side-effect of her public recognition.

In athletic events, the participants' performance is measured objectively and, therefore, winners are determined fairly. However, in work settings it's usually impossible to assess everyone's safety-relevant behavior objectively and to obtain a fair ranking for individual recognition. Hence, praising one individual in

public may lead to perceptions of inequality or favoritism from individuals who feel they did equally well, but did not get praised. Plus, such ranking sets up a win-lose atmosphere, appropriate for some sporting events but not in settings where teamwork is needed to achieve group goals and promote an interdependent injury-free culture.

It's beneficial, of course, to recognize *teams* of workers for their accomplishments, and this can be done in a group setting. Since individual responsibility is diffused or dispersed across the group, there is minimal risk of individual embarrassment or later peer harassment.

However, it's important to realize that group achievement is rarely the result of equal input from all team members. Some take the lead and work harder, while others loaf and count on the group effort to make them look good. Thus, it's important to deliver personal and private recognition to those individuals who go beyond the call of duty for the sake of their team.

### Let it sink in

In this fast-paced age of trying to do more with less, we try to communicate as much as possible when we finally get in touch with a busy person. After recognizing an individual's special AC4P effort, we are tempted to tag on a bunch of unrelated statements, even a request for additional behavior. This comes across as, "I appreciate what you've done, but I need more." Resist the temptation to do more than praise the safety-related behavior you saw. If you have additional points to discuss, it's best to reconnect later, after the supportive feedback has had a chance to sink in and become part of the individual's self-talk for self-recognition and self-motivation.

By giving supportive recognition, we give people a script they can use to reward their own behavior. In other words, supportive feedback strengthens the other person's self-reward system. Positive self-talk (or self-recognition) is critical for long-term maintenance of desired behavior. Thus, by allowing supportive feedback to stand alone and sink in, we enable the internalization of rewarding words that can be used later for self-motivation to perform the recognized behavior again.

**Use tangibles for symbolic value**

Tangible rewards can detract from the self-motivation aspect of quality recognition. If the focus of a safety-recognition process is placed on a material reward, the words of appreciation can seem less significant. In turn, the beneficial impact on one's self-motivation is lessened.

On the other hand, tangible rewards can add to the quality of interpersonal recognition if they are delivered as tokens of appreciation. Rewards that include a relevant AC4P safety slogan, as depicted below, can help to promote the desired behavior. But how you deliver a tangible reward will determine whether it adds to or subtracts from the long-term benefit of your praise.

The benefit of interpersonal recognition is weakened if the tangible consequence is viewed as a payoff for the safety-related behavior. However, if the reward is seen as symbolic of going beyond the call of duty for another person's safety or well-being, it strengthens the praise. Have the courage to tell it like it is: The AC4P wristband or another tangible reward is a token of appreciation and/or a symbol of going beyond the call.

**Consider secondhand recognition**

Up to this point, we've been discussing one-on-one verbal communication in which one person recognizes another for a particular AC4P behavior. It's also possible to recognize a person's outstanding efforts indirectly, and such an approach can have special advantages. Suppose you overhear a colleague talk to another person about your outstanding presentation regarding certain safe operating procedures. How will this secondhand recognition affect you? Will you believe the words of praise you overheard were genuine?

Sometimes people are suspicious of the genuineness of praise when it's delivered face-to-face. Is there an ulterior motive? Perhaps a favor is expected in return, as shown in the illustration below. Or maybe the recognition is seen merely as an extension of a communication exercise and thus devalued as sincere appreciation. Secondhand recognition, however, is not as easily tainted with these potential biases.

Suppose a colleague tells you that someone else told him about the superb job you did leading a certain safety meeting. What will be the impact of this type of secondhand recognition? Chances are you'll consider the recognition authentic because your colleague was only reporting what someone else said. Because that person reported your success to your colleague rather than you, there was no ulterior motive for the indirect praise.

Such secondhand recognition can also build a sense of belongingness or group cohesion among individuals. When you learn that someone was bragging about your behavior, your sense of friendship with that person will likely increase. When we talk about the achievement of others in behavior-specific terms, we begin a cycle of positive communication that can support desired behavior as well as activate self-talk for self-recognition and self-motivation.

Have the courage to initiate this cycle of positivism. This also sets an example for the kind of interpersonal communication that enhances self-esteem, self-efficacy, personal control, optimism, and group cohesion. As explained earlier, these are the very person-states that increase the potential for the occurrence of AC4P behavior and the achievement of an interdependent culture of compassion.[54]

### Accept supportive feedback

Most of us get so little supportive feedback from others that we are caught completely off guard when acknowledged for our commendable actions. We don't know how to accept recognition when we finally get it. Don't shy away when it does come; have the courage to embrace it. Remember the basic behavioral science principle: Consequences influence the behaviors they follow.

As the first lesson explained, positive consequences increase the probability the behavior recognized will continue. Plus, one's reaction to the supportive feedback influences whether the observer will attempt to provide supportive feedback again. Thus, it's crucial to react appropriately when receiving recognition from others. Let's consider seven basic guidelines for receiving supportive behavioral feedback.

### Don't deny or disclaim

Often when we attempt to give supportive recognition, we get a reaction that implies we're wasting our time. We get disclaimer statements such as "it really was nothing special" or "just doing my job." The most common reply is "No problem." This implies the behavior recognized was not special and did not warrant supportive feedback.

Supportive feedback needs to be accepted without denial or disclaimer statements, and without deflecting the credit to others. It's okay to show pride in your small-win accomplishments, even if others contributed to the successful outcome. After all, in an injury-free culture everyone goes beyond the call of duty on behalf of the well-being of others. Therefore, in this context, numerous people deserve recognition daily.

The truth is, supportive feedback will be intermittent at best for everyone. So when your turn comes, accept the positive feedback for your recognized behavior and for the many prior safety-supportive behaviors you have performed that went unnoticed.

Keep in mind that your genuine appreciation of the supportive feedback will increase the likelihood that the person who gave you feedback will give more behavioral recognition to others. Plus, you might be inspired to do the same, given that you realize personal benefit from the feedback and want others to experience such behavioral support.

**Listen actively**

Listen actively to the person recognizing you with supportive feedback. Certainly, you want to learn what you did right. Plus, you can evaluate whether the supportive feedback is given well. If the feedback does not pinpoint a particular behavior, you might ask the person, "What did I do to deserve this?" This will help to improve that person's method of giving supportive feedback in the future.

Of course, it's important to show genuine appreciation for the special attention. Consider how difficult it is for many people to go out of their way to recognize others with supportive feedback. So, revel in the fact that you're receiving some behavioral recognition, even if its quality could be improved.

**Use it later for self-motivation**

Most of your safety-related behaviors go unnoticed and unrecognized. You perform many of these behaviors when no one else is around to observe you. Even when other people are available, they are usually so preoccupied with their own routines they don't notice your extra effort for safety. So when you finally do receive supportive feedback for safety-related behavior, take it in as well-deserved.

Don't hesitate to relive this moment later by talking to yourself. Such self-recognition can motivate you to continue going beyond the call of duty on behalf of the safety of others. As mentioned earlier, self-talk can help you muster the self-motivation to perform more of the recognized behavior. It also contributes to holding safety as a value that remains preeminent, even when priorities change.

**Show sincere appreciation**

You need to show sincere gratitude with a smile, a thank you, and perhaps special words like, "You've made my day." Your reaction to receiving supportive feedback can determine whether that person will deliver similar recognition again. So be prepared to offer a sincere thank you, as well as words that reflect your pleasure in the memorable positive interaction. Plus, consider the courage the other person might have needed to give you that personal recognition.

We find it natural to add "You've made my day" to the thank you because it's the truth. When people go out of their way to offer us quality recognition, they *have* made our day. In fact, we typically relive such situations to improve a later day.

**Reward the recognizer**

When you accept supportive feedback well, you reward the person for his or her appreciating behavior. This can motivate that individual to deliver more positive feedback, especially if the person is an introvert and requires courage to step out and speak up to give interpersonal recognition.

Sometimes, you can even do more to assure the occurrence of more supportive feedback. Specifically, you can recognize the person for recognizing you. You might say, for example, "I really appreciate you noticing my safety-relevant behavior and calling me a leader of the AC4P process to sustain an injury-free workplace." Such supportive and rewarding feedback provides direction and motivation for those aspects of occupational safety that are especially worthwhile and need to become routine.

**Ask for recognition**

If you feel you deserve supportive feedback, why not ask for it? You will likely view this recognition as less genuine than if it had been spontaneous, but the outcome from such a request can be quite beneficial. You might even receive some words worth reliving later for self-motivation. Most important, you will remind the other individual in a nice way that a prime opportunity to offer supportive feedback was missed. This could be a valuable learning experience for that person.

Consider the possible beneficial impact from your statement to another person that you are pleased with a certain outcome of your extra effort. With the right tone and affect, such verbal behavior will not seem like bragging but rather a declaration of personal pride in a small-win accomplishment—something more people should feel and relive for self-motivation. The other person will surely support your personal praise with supportive testimony, and this will bolster your self-motivation. Plus, you will teach the other person how to support the commendable behavior of others.

## From Principles to Practice

William James, the first renowned American psychologist, wrote, "The deepest principle in human nature is the craving to be appreciated."[55] Then in 1936, Dale Carnegie advocated the key to winning friends and influencing people is to "always make the other person feel important."[56] How can we readily fulfill the human need to feel appreciated and important? The answer, of course: Give and receive supportive feedback well.

Before applying the interpersonal feedback techniques discussed here in the field, we suggest participants role-play the procedures for delivering and receiving supportive and corrective feedback. Such role-play should be followed with systematic behavioral feedback from the instructor and other participants.

The delivery and reception of this feedback should follow the guidelines presented here, thereby providing observational learning for the participants. In addition, we recommend the participants practice the recommended feedback techniques on family members and/or good friends before taking their behavior-improvement techniques to the workplace. This allows the participants to learn where they can best improve their communication skills and understand their best method in delivering a message.

Also, we suggest the participants work on mastering supportive feedback before attempting corrective feedback. Supportive feedback, or behavioral recognition, is much easier to employ than corrective feedback, and can do no harm if applied incorrectly. On the other hand, an inappropriate application of corrective feedback could result in a negative attitude or hurt feelings and loss of interpersonal trust that could hinder future constructive interactions between these colleagues.

After discussing the benefits of providing supportive feedback, the instructor should review the basic delivery and receiving steps for effective supportive feedback, as detailed above. Then, the participants should pair off and practice giving each other supportive feedback for a safety-related behavior they observed. One participant should agree to be the feedback *sender* and the other is the *recipient*. Then, a particular work situation and safety-related behavior should be defined for the role-play exercise. This could be a real-world situation the participants have experienced or a scenario that is likely to occur in the future.

After receiving the supportive feedback, the recipients should express their reaction to the feedback. Did the feedback seem genuine? Did the feedback address a specific behavior? Did the recipient feel rewarded by the interchange? Afterward, the roles should be reversed, with the *sender* playing the role of feedback *recipient* and vice versa.

The instructor/facilitator should circulate among the practicing role-players and note examples of particularly effective and ineffective performance. These observations could then be discussed when the participants reconvene as a group. Group members should discuss their feedback experiences from the perspectives of both the *sender* and the *recipient* of supportive feedback.

Group leaders might select a feedback pair who displayed exemplary performance during the role-play sessions and ask these participants to demonstrate their interaction skills to the rest of the group. The group facilitator might offer supportive and corrective feedback to the presenters regarding their delivery and receipt of interpersonal supportive feedback, and then ask other participants to contribute supportive and corrective feedback.

After role-playing and gaining confidence in your ability to catch safety-related behavior and give genuine supportive feedback, you are ready to reap the benefits of real-world experience. Observe the ongoing behaviors of certain individuals (perhaps starting with family members). After noting a good behavior, apply the supportive feedback process. Record your experiences on the Documentation of Delivering Supportive Feedback for Safety-Related Behavior form on the next page. After completing this, it would be instructive to discuss the information (especially the recipients' reactions) at group meetings.

**Documentation of Delivering Supportive Feedback for Safety-Related Behavior**

| Behavior Recognized | Environmental Setting | Reactions of Recipient |
|---|---|---|
| 1. | | |
| 2. | | |
| 3. | | |
| 4. | | |
| 5. | | |

Role-playing, group demonstrations, and interactive feedback should also be used to improve skills of the participants at giving *corrective* feedback. Practice sessions should follow the same basic format as that given above for *supportive* feedback, except that it will take more practice and individual direction to master this type of verbal communication.

After participants demonstrate with role-playing that they can follow the basic steps of delivering corrective feedback, they are ready to apply this new communication skill beyond the classroom. Again, we recommend the first few corrective feedback sessions be attempted with family members or friends after observing an undesired behavior (of course).

Subsequently, at-risk behavior observed among work colleagues should be the target of the corrective feedback communication. For maximum benefit, these sessions should be documented in the corrective feedback form on the following page and discussed in group meetings among those participating in this ongoing exercise.

**Documentation of Real-World Corrective Feedback for At-Risk Behavior**

| Undesirable Behavior | Environmental Setting | Alternative Behavior | Recipeint's Reaction |
|---|---|---|---|
| 1. | | | |
| 2. | | | |
| 3. | | | |
| 4. | | | |
| 5. | | | |

Our workshop participants have reported that the most challenging part of giving corrective feedback is the opening statement. You see someone working at risk, and the need for corrective feedback is obvious. If the behavior is putting the person in obvious imminent danger, a stop warning is natural and likely accepted. But usually this is not the case. For example, you see an individual driving a vehicle without using the safety belt, or a coworker just climbed a flight of stairs without holding a handrail. What would you say? How could you set the stage for corrective feedback without offending this individual or turning him or her off to listening to you further?

As we mentioned earlier, you don't want to insult people and perhaps lower their self-esteem or self-efficacy. You need to take an empathic stance, as discussed in the next lesson, and avoid words that challenge knowledge or competence. For example, as my friend John Drebinger suggests in his book *Would You Watch Out For My Safety?*, you might start with, "As you know," before pointing out an at-risk behavior that is ongoing or has just occurred.

Alternatively, you could start a corrective feedback conversation with, "Hey, could I actively care for your safety?" Few would say no to this request, especially if AC4P behavior is part of the mission statement for your organization. So, a yes answer to your AC4P question gets your foot in the door and gives you overt permission to offer corrective feedback. And then you might precede your behavior-based advice with something like, "As you know. . . ."

## Discussion Questions

1. **It would be useful to discuss this challenge of starting off a corrective feedback conversation with words that invite rather than stifle an AC4P corrective-feedback conversation. Write down here some examples of phrases you could use. It is likely participants will offer alternative opening dialogue than the words suggested here.**

2. **Finally, it would be very useful to practice and document occasions of giving feed*forward* and feed*back* to colleagues throughout your day. You should note here what went well and not so well in your delivery of personal *feedback* for behavioral support or correction. Reflect on these comments as personal *feedforward* to improve your next opportunity to improve another person's behavior.**

# Lesson 5

## EMBRACE AND PRACTICE EMPATHY

The rationale for using more supportive than corrective feedback to improve behavior is based on the different feeling states provoked by positive versus negative consequences. Similarly, the way a behavior-improvement process is implemented can increase or decrease feelings of empowerment, build or destroy trust, and cultivate or inhibit a sense of teamwork or belonging.

Decisions regarding which behavior-improvement process to implement and how to refine existing intervention procedures should be based on both objective observations of behaviors and subjective evaluations of feeling states. Often, it's possible to employ empathy to evaluate the indirect internal impact of an intervention. Simply imagine yourself in a similar situation and then ask yourself, "How would I feel?"

Empathy is not the same as sympathy, although dictionary definitions are similar. The *New Merriam Webster Dictionary* defines sympathy as "the capacity for entering into and sharing the feelings or interests of another"[57] and empathy as "the capacity for experiencing as one's own the feelings of another."[58]

Likewise, *The American Heritage Dictionary* defines empathy as "identification with and understanding of another's situation, feeling, and motives,"[59] in contrast to sympathy as "a feeling or expression of pity or sorrow for the distress of another person."[60] Thus, we sympathize when we express concern or understanding for another individual's situation, but we empathize when we identify with another person's situation and realize what it's like to be in the other person's shoes.

An empathic level of awareness and appreciation is not easy to achieve and can only be reached after we minimize the reactive filters that bias our

conversations and then listen intently and proactively to another person. Not only must we hear every word, but we must also look for feelings, passion, and commitment reflected as much in body language and manner of expression as in the words themselves. Begin with the mind-set that there's more than one side to every story. The other person's perspective might be very different than yours, as reflected in the illustration above.

When we observe another person's behavior, we should try and view the situation from that individual's perspective. When we listen to excuses for at-risk behavior, we can try and see ourselves in the same predicament. We can imagine what defense mechanisms we might use to protect our own ego or self-esteem. And when we consider action plans for improvement, we should try and view various alternatives through the eyes of the other person.

Are you thinking it's difficult to see situations and circumstances through the eyes of another person? Are you thinking, "This is easier said than done"? Well, you're probably right. But consider that we're only talking about taking a different perspective into our conversations. We need to approach our coaching conversations with an empathic mind-set.

## Empathic AC4P Listening

Sometimes, in their eagerness to make things happen, safety professionals, supervisors, teachers, coaches, and parents give feedforward directions and corrective feedback in a top-down, seemingly controlling manner. In other words, the passion to make a difference can lead to an overly directive approach to getting others to change their behavior. However, an indirect or nondirective approach to giving advice is usually more effective, especially over the long term. This is a basic tenant of humanistic therapy.

Think about it: How do you respond when someone tells you exactly what to do? Now it certainly depends on who is giving the directive, but I bet your reaction is not entirely positive. You might follow the instructions, especially if it comes from someone with the power to control consequences. But how will you feel? Will you be self-motivated to make a lasting change? You might be if you asked for the direction. But if you didn't request advice or feedback, you could feel insulted, embarrassed, and/or disrespected. Try to be more nondirective when using interpersonal conversation to affect behavior change. This requires empathic listening.

Dale Carnegie wrote about the value of empathic listening more than eighty years ago in his classic book, *How to Win Friends and Influence People*. His wisdom is reflected in the writing of many authors of popular self-help books, including Stephen Covey and his fifth habit of highly effective people, "Seek First to Understand, Then to Be Understood."[61]

Carnegie, Covey, and others offer the same basic strategies for empathic listening, and if you've had any training in effective communication, you've heard the same advice. Let's review these guidelines with four easy-to-remember words, each beginning with the letter *R*.

For readers who have received communication training, this review will at least provide a mnemonic for remembering how to listen with an empathic AC4P mind-set and teach others to do the same. To be sure, the increasing "lean and mean" and win/lose paradigms of contemporary organizations, as well as the focus on impersonal emails and text messages, suggest a dire need to teach and use these humanistic guidelines for one-on-one empathic communication.

## Repeat

This is the easiest technique to use; simply mimic (or repeat) what you heard with the same words. This clarifies that you heard correctly and, most important, prompts the person to say more. Remember the purpose of empathic listening is to motivate the other person to say more so you can truly understand the problem.

So, if a friend tells you he's quitting his job, you might repeat this statement with, "You're quitting?" This shows you're attentive, listening, interested, and waiting for more information. Hearing how drastic the statement sounds, the person might reply, "Well, at least I feel like quitting." Then, what would you say? Following this repeat technique, you would say, "You mean you feel like quitting?" Or, you might use different words to echo the same meaning. This is the next empathic listening technique.

**Rephrase**

Instead of mimicking the content, you might rephrase the statement, saying back the same thing but in different words. In our example, you might say, "You mean you don't like working here anymore?" By putting the statement in your own words, you're showing genuine interest while also asking for more information.

You're also checking for understanding. If you can rephrase the statement correctly, you have received and interpreted the communication accurately. If you have not heard the information correctly, the person might then repeat the statement. It's possible your friend miscommunicated or you misperceived something. Your rephrasing gives the other person a chance to explain. And, this is what you want—more disclosure of the problem. Suppose your friend clarifies, "Well, it's not that I don't like working at this job site, it's just that the work supervisor gets me so frustrated at times, I feel like quitting."

Now, your friend has revealed a more specific issue and you return with, "Your supervisor gets you frustrated." Alternatively, you could attempt to rephrase with something like, "You mean your work supervisor gets you so angry that your motivation to continue working here is sapped." Or, perhaps this statement calls for the next *R* of empathic listening: *Ratify*.

**Ratify**

With this listening strategy you demonstrate affirmations or support for the individual's statement by confirming your understanding. You offer words that show approval for what is being said, and this in turn encourages more explanation. In this example, you might ratify your appreciation of the statement by saying, "I know the feeling. I've been frustrated with my work supervisor at times and wonder whether this job is for me." This also makes it personal by showing you have had the same feelings.

At this point, you might be tempted to jump in with probing questions to find out more about the frustration, the supervisor, or the situation. When did that supervisor get you so upset? What did he say or do? Why are you so frustrated that you want to quit? But wait. Resist the temptation to be directive.

You probably have not heard enough about the problem to begin a structured and unbiased analysis.

More empathic listening could reveal problems beyond a particular supervisor. Perhaps it's not the supervisor per se but a particular work assignment or unreasonable deadline. Or the problem might stem from interactions with a coworker, or a family member, or from feelings of personal inadequacy, including a perceived loss of confidence, self-efficacy, or personal control.

A person's distress signals can come from many sources, and these will probably not come to the surface quickly in one-on-one communication. And even if the relevant causes of a problem are disclosed early in the conversation, it's unlikely you could give optimal advice or directive action at this point that is both useful and accepted.

Usually, the best we can do is listen actively with repeat, rephrase, and ratify strategies in order to get the problem out in the open. Ultimately, we want the person to express true feelings, as indicated by the fourth *R*.

### Reflect

When people reflect on their inner feelings about a situation, they are at the personal root of the problem. Such self-disclosure of person-states can lead to insight into the true cause of the problem—for both the speaker and the listener—and suggest strategies for intervention. However, even at this stage (with outer layers of the onion peeled away), it's usually better to let the speaker entertain a variety of possible intervention approaches.

If you've been an empathic listener, you might eventually get the ultimate reward for your sensitivity, patience, and emotional intelligence. The speaker will ask you for specific advice. When you hear words like "What do you think I should do," you have mastered empathic listening. You have shown you actively care, and now your thoughtful direction will likely be most relevant, understood, and accepted.

## Ask Questions First

Suppose the conversation is not about a serious issue such as personal distress, frustration, or apathy, but about a less-than-desired behavior. You see an opportunity for a person to show more or better AC4P behavior for safety in a particular situation. What should you say?

Instead of telling the person what to do, try this. Get the individual to tell you, in his or her own words, what he or she could have done to be more effective from an AC4P mind-set. You can do this by asking questions with a sincere and empathic demeanor. Avoid a sarcastic or demeaning tone at all costs.

First, point out certain desirable safety-related behaviors you noticed—it's important to start with positives. Then move on to the less desirable behavior by asking, "Could you have been more effective in that situation?" Of course, you hope for more than a yes or no response to a question like this. But if that's all you get, you need to be more precise in follow-up questioning.

You might, for example, point out a particular situation where the work could have been accomplished more safely and then ask what behaviors could have been safer. Here we're talking about giving corrective feedback, as was addressed in the prior lesson. Now you realize the value in starting corrective feedback with a question. For example, you could ask, "Are those the correct cut-resistant gloves for the task?" or "Why was the safety guard on that grinder removed?"

By asking questions, you're always going to learn something. If nothing else, you'll hear the rationale behind the at-risk behavior. You might uncover a barrier to the safe behavior that you can help the person overcome. A conversation that entertains ways to remove obstacles that hinder safe behavior is especially valuable if it translates possibilities into feasible and relevant action plans.

Obviously, empathic listening, diagnosing, and action planning take patience. Conversations at this level are often not efficient, but they are always most effective. Through questioning and listening, the objective is to first learn what it's like to be in the other person's shoes. Then the objective shifts to developing a corrective approach that fits the circumstances as mutually understood by everyone involved in the conversation. If commitment to follow through with a specific action plan is stated, you were an AC4P behavioral coach.

## From Principles to Practice

Okay, we've reviewed the rationale for this empathy lesson and offered guidelines for accomplishing empathic listening. You've been educated. You understand the value and the purpose of empathic listening. Can you be an effective empathic listener? Of course you can, but it may take some practice, along with relevant behavioral feedback. Now we're talking about training. We suggest the following group exercise:

Divide the participants into pairs and ask them to find a relatively quiet location for a brief conversation. If the size of the group and/or the facility prevents private dyadic communication, simply ask participants to turn to the individual closest to them and execute this practice/feedback session on the spot. While focusing on their private one-to-one conversations, the participants will readily tune out the verbal noise from other dyads.

If time permits, tell participants this exercise has two trials. For the first trial, one member of the dyad will serve as the first speaker, and the other will be the empathic listener. Then for the second trial, the dyadic partners will switch roles.

For each trial, the speaker should tell the listener a personal story, perhaps one that could benefit from another person's perspective and/or advice. Emphasize that the speakers will have a maximum of three minutes to tell their story while the listeners practice the four guidelines for empathic listening: Repeat, Rephrase, Ratify, and Reflect.

After the three-minute story, the listeners should take a maximum of two minutes to offer their viewpoints and advice (if relevant) while the former speakers play the role of empathic listeners. Afterwards, have each member of the dyad offer behavioral feedback regarding the demonstration of effective versus ineffective listening skills.

Start with the positives first (i.e., supportive feedback) and then offer some corrective feedback. The challenge: Practice the recommended ways of giving and receiving supportive and corrective feedback.

When considering the nature of feedback to give the listener, ask yourself these questions:

**1. How did the listener facilitate continued verbal behavior from the speaker?**

2. **What did the listener do to encourage the speaker to reveal personal aspects of his or her story?**

3. **Did the speaker ask for advice or did the listener offer advice without a request?**

4. **Was the advice relevant and/or useful?**

5. **What particular R-word guidelines for empathic listening were used in this dyadic communication?**

**6. What could the listener have done to facilitate more revealing communication from the speaker?**

**7. How did this exercise help you appreciate the value of empathic listening, and perhaps realize validity in the earlier statement, "It's not that easy"?**

The instructor should call time after the three-minute story and also after the two-minute reaction by the listener. Then allow five minutes of feedback discussion among the dyads. Total time for Trial 1: ten minutes. If time permits, this entire exercise should be repeated with the speakers and listeners switching roles (i.e., Trial 2).

After the dyads complete one or two trials, the instructor should facilitate a group discussion about the takeaways from this exercise. Ask the participants to report what they learned about empathic listening from their discussions.

**What R-word strategies did they notice, and what were advantages, if any, of practicing these listening techniques?**

**In what ways, if any, did this experience influence your intentions (i.e., feedforward) for subsequent safety-related communication with work colleagues or family members? Note: As you know, when people make a public commitment to improve, the probability of actual improvement increases markedly.**

# Lesson 6

## DISTINGUISH BETWEEN MANAGING BEHAVIOR AND LEADING PEOPLE

Management is not the same as leadership. Yes, both are critically important for bringing the best out of people. Simply put, managers hold us accountable to perform desirable behavior and avoid undesirable behavior. Leaders, however, inspire us to hold ourselves accountable to do the right thing.

Managers control behavior with an external (or extrinsic) accountability intervention or system. Leaders facilitate self-motivation by influencing person-states (e.g., perceptions, attitudes, and/or emotions) that facilitate self-motivation. Self-motivation (or self-direction) often leads to discretionary behavior, behavior beyond that which is required. To be sure, safety managers can be safety leaders.

### Safety Leaders Focus on Process

Safety managers are typically held accountable for outcome numbers. They use outcome numbers to direct the behavior of those who report to them. Most followers or subordinates of managers are assigned their subordinate responsibilities, and they did not choose their manager. In safety, outcome numbers are based on the relatively rare occurrence of an injury. These numbers (e.g., total recordable injury rate [TRIR]) are reactive, reflect failure, and are not diagnostic for injury prevention.

In contrast, safety leaders hold people accountable for accomplishing proactive process activities that can prevent harm and lower injury rates. When improvement in process activities is observed, leaders provide those responsible with positive recognition for their efforts. Those who feel appreciated for their safety proactivity develop a sense of personal responsibility for continuing to make safety-related contributions.

## Safety Leaders Educate

In occupational safety, training is more common than education. Managers want employees to know exactly what they need to do in order to complete a particular task effectively and safely. With a training mind-set, however, managers can come across as demanding: "Do this because I said so."

As discussed earlier, education involves explanation. The principles or the rationale behind a particular set of procedures are elucidated for employees. Education answers the *why* question—why a certain protocol needs to be followed. By extension, it also answers the critical question, "What's in it for me?" By taking the time to explain rules and procedures, effective educators help people develop self-accountability for a safety action plan rather than doing something a certain way because a manager or supervisor is holding them accountable.

Proper education of principles underlying a safety process can facilitate creative customization and personal ownership. When leaders explain a rationale and offer examples rather than issue a policy or top-down directive, individuals and teams can select work and safety procedures that best fit their situation. Throughout the process of refining a set of procedures, employees assume ownership and follow through. They become self-directed rather than other-directed.[39]

## Safety Leaders Listen First

Managers often speak first and later listen to concerns or complaints. This reflects the typical top-down urgency to get a job done and is a reasonable strategy for efficient productivity. After all, managers must make things happen according to a prearranged plan, implementing explicit directives and a mechanism for ensuring compliance. Managers describe an action plan and the accompanying accountability system, and then they answer questions from workers who want to be sure they can do the right thing.

In contrast, leaders first take the time to learn another person's perspective before offering direction, advice, and/or support. They take the time to recognize the other experts in the room and the opinions and advice they might have to offer. As explained in the previous lesson, this is empathic listening, and it is vital to diagnosing a work-related situation before urging change or continuous improvement

This is certainly not the most efficient way to complete a work assignment. It requires time, patience, and empathic communication. Many questions are asked before giving advice. The payoff: An individual or team willingly and creatively customizes an action plan or process to achieve a particular safety-related outcome.

## Safety Leaders Promote Ownership

When leaders give a reasonable rationale for a desired outcome and offer opportunities for employees to customize methods to achieve that outcome, they facilitate self-directed motivation. This is commonly referred to as self-motivation, as explained in detail later in this lesson. With this motivational mind-set, employees participate because they want to, not because they have to. This is discretionary behavior.

Managers who direct by command might get compliance with an action plan, but they might also stifle self-accountability or self-motivation. Compliant behaviors—those following a prescribed standard, policy, or mandate—are other-directed. Work is performed and safety protocols are followed to satisfy someone else.

These behaviors are likely to cease when that *someone else* is not monitoring them—when a manger is not holding them accountable. A common result: Safety-related behavior (e.g., using PPE) is practiced at work but not at home where individuals' self-directed actions are not monitored.

## Safety Leaders Set Expectations

Most behavior starts as other-directed. It's performed because it was requested or mandated by another person. Does this behavior remain other-directed, or does it evolve to a self-directed or self-motivated state? How you ask for the performance outcome you desire can be a game changer. A behavioral request that comes across as a mandate is likely to remain stuck as other-directed. This is often how many managers approach safety. Just look at the traditional emphasis on regulatory compliance and the all-too-common slogan "Safety is a condition of employment."

Leaders facilitate a shift from other-directed to self-directed motivation. How? They initiate a process or action plan with expectations rather than mandates. Expectations imply choice. A certain outcome can be anticipated, but expectations allow room for individual and group decision-making regarding the procedures and methods to reach that outcome.

When employees realize what's expected of them—but are given personal control in setting goals and determining how to reach them—they are more likely to own the process. Thus, leaders facilitate a transition from an other-directed to a self-directed mind-set.

## Safety Leaders Look beyond the Numbers

Managers focus on numbers. In safety, this means injury records and workers' compensation costs. When we have discussed BBS, PBS, and AC4P principles and procedures with managers, inevitably the question arises, "What's the ROI [return on investment]?" Managers focus on how much a process will cost and how long it will take for the numbers (as in TRIR) to improve. This approach is reflected in the popular management maxim "You can only manage what you can measure."

This numbers-based analytical mind-set has caused BBS to become much less effective than it could be. All too often, the focus of BBS is on numbers. Managers want to see the number of behavioral observations, and thus the frequency of completed checklists increases. As mentioned earlier, this ignores the real purpose of a proactive AC4P safety process: to engage employees in meaningful behavior-focused communication with each other about how to work more safely and achieve a vision of "zero harm."[18]

A limiting BBS focus on collecting numbers for a computer database and a statistical analysis leads to repeated use of the same behavioral checklist (i.e., CBC) per work area, and then many workers pencil whip checklist completion by only marking the boxes but avoiding the most important part of the process—interpersonal, supportive, and corrective behavioral feedback. The critical purpose of the behavioral observation-and-feedback process—an AC4P conversation—is missing.

Managers often celebrate invalid and inflated percent-safe scores rather than recognize the more important purpose of AC4P coaching: workers giving each other on-the-job behavioral feedforward and feedback for proactive injury prevention.

As stated above, no matter how you crunch the numbers, behavioral observations will never produce statistically reliable data. Are these numbers then useless? Of course not. The behavioral data pinpoint at-risk behaviors performed at certain job sites. And if the comment section on the CBC is used effectively, environmental and/or management factors that contribute to these at-risk behaviors are revealed. If this information is communicated to the appropriate personnel, positive change should be expected.

Safety leaders surely appreciate the need to hold people accountable with numbers, but they also understand you can't measure everything. Sometimes you do things and ask others to do the same because they are the right things to do.[62]

Safety leaders understand it's important to increase person-states you can't readily put a number on—self-esteem, self-efficacy, personal control, optimism, and a sense of belonging. As discussed earlier, these five intangibles influence people's tendency to go beyond the call of duty on behalf of another person's safety, health, or well-being—to perform AC4P behavior.

Safety leaders do things on a regular basis to inspire these person-states in employees throughout a work culture. They don't worry about measuring their own direct impact on these intangibles. They have confidence in the research that indicates promoting these five states is important for increasing one's propensity to perform AC4P behavior.[24] Analogously, people take vitamin pills daily even though they don't notice any direct measurable impact on how they feel.

It's a good idea to occasionally assess whether certain actions influence people's subjective feelings or person-states in desirable directions. This can be done informally through personal interviews, unaided by a score card. It's also a given that certain interpersonal and group activities are useful.

Genuine one-to-one recognition increases self-esteem and self-efficacy; behavior-based goal setting builds perceptions of personal control and optimism; and group celebrations facilitate a sense of belonging or interpersonal relatedness. Safety leaders perform and support these activities with passion and patience, realizing an improvement in the numbers of an accountability system might not be immediate.

Safety leaders don't need a monitoring scheme to facilitate their attempts to help people feel valuable and part of an important team effort. They are self-directed and self-motivated, and this helps inspire self-motivation in others. Safety leaders set an AC4P example, knowing full well the positive impact it has on cultivating an injury-free work culture.

Bottom line: Safety management by policy, directive, and regulation is necessary at times to motivate people to do the right things for injury prevention. But this approach alone will not achieve an injury-free workplace. Safety managers need to know how and when to step up and become safety leaders.

Here is the most important takeaway: You don't have to occupy a safety management position to be a safety leader. By following the seven lessons of this teaching/learning manual, you can help people transition from an other-directed to a self-directed or self-motivated person-state. Let's delve deeper into this desirable person-state of self-motivation.

## Self-Motivation

The three *C*s of self-motivation—Choice, Competence, and Community—illustrate the three evidence-based perceptions or person-states that determine self-motivation.[29] Dispositional, interpersonal, and environmental conditions that enhance these states increase personal perceptions of self-motivation. Consider, for example, how proper application of the first five lessons can increase one's perception of competence and hence fuel self-motivation. Consider, also, how our language can affect each of these perceptions.

### Watch your language

Your language should suggest minimal external pressure. The common phrases "Safety is a condition of employment," "All accidents are preventable," and "Bullying is a rite of passage" reduce one's sense of autonomy. In contrast, the slogan "Actively caring is a core value of our organization" implies personal authenticity, interpersonal relatedness, and human interaction.

The common phrase "random acts of kindness" has a disadvantage when describing or promoting AC4P behavior. Random implies the behavior happens by chance, suggesting it's beyond individual choice or control. The kind act may appear random to the recipient, but it was likely performed intentionally and was self-motivated. An alternative and more accurate phrase is "mindful acts of kindness." The language we use to prescribe or describe behavior influences our perceptions of its meaningfulness and its relevance to our lives. Language impacts culture, and vice versa.

## Opportunities for choice

Participative management means employees have personal choice during the planning, execution, and evaluation of their jobs. People have a need for autonomy, regardless of dispositional and situational factors. In the workplace, managers often tell people what to do rather than involving them in the decision-making process. Referring to language again, should managers give mandates or set expectations? Should they demand compliance or ask for commitment?

In the workplace, employees are often viewed as passive followers of safety rules and regulations. Why? Because mangers plan and evaluate most aspects of the job, including the safety protocol. The wage worker's perception of choice is limited. Yet, as we've stressed throughout this teaching/learning manual, an injury-free workplace requires interdependent engagement, information gathering, and AC4P coaching by the line workers, who are the individuals who know most about the hazards and at-risk behaviors and the factors contributing to these potential determinants of injuries and fatalities.

## Involve the followers

Autonomy and the perception of choice are supported when rules are established by soliciting input from those affected by the regulation. Employees are more likely to comply with regulations they helped to define. Shouldn't they have substantial influence during the development of safety policy they will be asked to follow? After all, line workers know best what actions should be avoided versus promoted in order to achieve and nurture an injury-free workplace.

What mechanisms are in place for wage workers to provide policy recommendations at your workplace? What safety-suggestion system is provided for safety-related input from employees? As you know from our prior discussion, such a system is considered feedforward.

Years ago, "safety suggestion boxes" were common in most companies, but many have been removed. Is this the case because we now have the Internet for more convenient and efficient input from employees? This may be true for some organizations, but our observations suggest another explanation: extinction. As discussed below in this lesson, desirable behavior stops occurring if it is not followed by a supportive consequence.

Numerous participants at our AC4P workshops have told us they once offered safety suggestions on paper (i.e., a "safety suggestion box") or on an Internet site but stopped this behavior because "no one paid attention." They

saw no evidence that their safety suggestions were considered. The illustration above reflects their perception regarding safety suggestions, and their rationale for not participating further in the organization's safety suggestion system. In the absence of supportive consequences, desirable behavior extinguishes.

**Support feedforward with feedback**

Do you want to boost engagement in a safety-suggestion system whereby workers routinely provide feedforward relevant for injury prevention? Of course you do, and you already know how to do it. Read each suggestion carefully and provide your employees with specific and timely feedback. A one-on-one conversation about this is best, but a professional and supportive feedback letter would also be greatly appreciated.

Yes, feedforward requires feedback. At least send an email response to those who offer a safety suggestion, stating your sincere appreciation for their AC4P support of safety. Specify any and all aspects of a feedforward suggestion that will be seriously considered for implementation. Imagine the self-motivation you will inspire when employees see a workplace application of their suggestion. These participants will surely tell others of their bottom-up influence of occupational safety, and this would surely encourage more employee involvement in the safety-suggestion system.

Even safety suggestions that are not feasible, often because they require unavailable resources, should be acknowledged with sincere and personal appreciation (certainly not a form letter). Offer a reasonable explanation for your decision and indicate that a thoughtful suggestion not used immediately will be retained for potential consideration in the future. With empathy, consider that this employee chose to offer feedforward for safety improvement. Your sincere and personal recognition will fuel this individual's self-motivation to do more for occupational safety and health.

### A memorable example

Two decades ago, SPS offered AC4P training and consulting at the Toyota Motor Manufacturing Plant in Georgetown, Kentucky. The company had incorporated the most remarkable and successful employee-suggestion system I (Scott) had witnessed at the time and have ever seen since. I remember vividly how the line workers showed me with pride many minor improvements in their work environment that resulted from their employee-suggestion system. Their feedforward process addressed more than safety—any idea relevant to improving the quality of safe manufacturing at this world-class facility was welcomed.

At the time, approximately six thousand workers were employed at this site. How many suggestions about work quality or safety would you expect to receive each month from these employees, given a work culture that actively promotes and supports an employee-suggestion system? Would you expect six thousand suggestions per month—an average of one per employee? The answer: Employees submitted more than twenty-five thousand suggestions per month in 1996.

Difficult to believe, yes, but let's consider how this feedforward system was supported. First, three employees spent their entire workday reading suggestions and preparing a customized feedback letter for each. Naturally, many suggestions could not be implemented, and several were redundant. But most contained some useful information for improving work quality and/or safety, and the feedback letters acknowledged any useful information.

While these supportive feedback letters certainly reinforced the submission of suggestions, there was a much larger potential reward. If a suggestion was implemented and reaped any financial benefits for Toyota, the individual(s) who offered the suggestion received 10% of any annual profit (or savings) the company realized from the feedforward.

Back in 1996, I was told this feedforward system cost Toyota Georgetown $1 million per year to implement. But the annual profit from this employee-based suggestion system was estimated at $5 million.

While the Toyota example may seem impractical for many companies, it does show the dramatic benefits that can come from a feedforward system that solicits continual advice from line workers. The continuous-improvement Kaizen spirit of the Toyota culture certainly facilitated participation. But the consequences of a feedback letter and a potential share of financial profits were certainly instrumental in motivating feedforward submissions.

A takeaway: Any feedforward system that actively solicits ongoing safety suggestions from workers at risk for an injury can provide invaluable information. But to succeed over the long term, such a system requires soon, certain, and positive consequences for the participants. A feedback letter should acknowledge sincere appreciation of each suggestion and specify how the feedforward advice will be applied, if only to stimulate future decision making.

## Empowerment

In the management literature, empowerment typically refers to delegating authority or responsibility, or to sharing decision-making. In other words, when managers say, "I empower you," they usually mean "Get 'er done." As reflected in the illustration below, the message is, "Make it happen; no questions asked." In contrast, the AC4P safety leader first assesses whether the empowered individual *feels* empowered. A person's proper assessment of feeling empowered involves asking three questions, as derived from social learning theory.

### Self-efficacy

As depicted in figure 6, the first question, "Can I do it?" asks whether the empowered individual or group has the resources, time, knowledge, and ability to handle the assignment. The knowledge and ability components refer to training, and the term *self-efficacy* places the focus on personal belief. An observer might think an individual has the competence to complete a task, but the empowered person might feel different. Thus, a yes answer to the first empowerment question implies a belief among those who received the assignment that they have the relevant personal competence to achieve the process or outcome goal.

### Response-efficacy

The second question, the response-efficacy question, asks whether those who are empowered believe pursuing and accomplishing the assignment or attaining the process goal (i.e., performing the required behavior effectively) will contribute to a valued mission of the organization, work team, or individual.

Regarding BBS, the response-efficacy question translates into believing the AC4P coaching process with the CBC will eventually contribute to injury prevention and perhaps help to cultivate an AC4P brother's/sister's keeper culture.

A sports team would answer yes to this question if the athletes believe their new workout routine or competition strategy will increase the probability

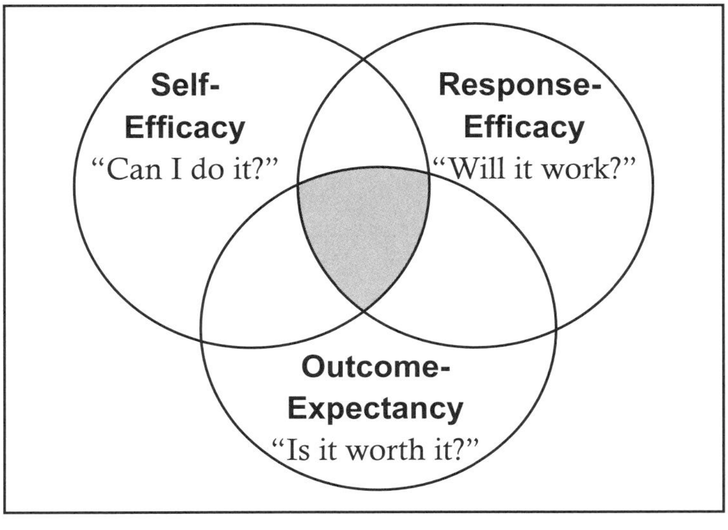


**Figure 6.** The three beliefs that determine the empowerment person-state

of winning a game. Students studying for an exam would give a yes answer to response-efficacy if they believe the study strategy will contribute to earning a higher exam grade. Of course, the behavioral outcome for these two examples could be more distal and substantive, like having a winning season or obtaining a college degree, respectively.

A negative answer to the self-efficacy question indicates a need for more training; a negative answer to the response-efficacy question implies a need for education. In other words, people might believe that they are able to accomplish a particular process or task (i.e., self-efficacy) but not believe such accomplishment will make a difference in the desired outcome (i.e., response-efficacy). In this case, education is needed, including an explanation of an evidence-based principle or theory and perhaps the presentation of convincing data. Regarding the AC4P behavioral coaching process, substantial data are available to show that an AC4P behavioral observation-and-feedback process prevents workplace injuries (e.g., see table 1 on p. 6).[2]

**Outcome-expectancy**

The third empowerment-assessment question targets motivation. Is the expected outcome worth the effort? The performance of relevant behavior is motivated by anticipating a positive consequence to achieve or a negative consequence to avoid. Referring back to our first lesson, recall that people feel more choice and are more likely to be self-motivated when they perceive they are trying to achieve a positive consequence than when they're responding to avoid a negative consequence.

When it comes to safety-supportive behavior, which of the three empowerment questions is most difficult to answer with an empathic yes: (1) Can I do it? (2) Will it work? or (3) Is it worth it? The answer to our question is obvious, right?

Consider the nationwide safety issue of texting while driving. First, we have self-efficacy. It's easy to put your smart phone away and choose not to text or answer the phone if it rings. Plus, it's likely you do believe the common-sense statistic that texting by drivers has caused numerous fatal car crashes. If every driver stopped texting while driving, vehicle crashes would decrease substantially, right? Thus, it's easy to answer yes to the response-efficacy question, Will it work? But self-efficacy and response-efficacy are not enough to stop many individuals from using a hand-held smart phone while driving.

And this is the case for most, if not all, safety-related behaviors, right? Most employees are trained to perform their job safely. The injury-prevention function of the safe behavior is usually obvious and is often supported with concurrent injury statistics. So, most workers know how to perform a task safely and believe the extra safety precautions will reduce the likelihood of an injury. Yet they often take risks, knowing the chances of a mishap and an injury are increased. Why, then, do they take risks?

Of course, you know the answer. It's that third empowerment question, the consequence or motivational domain: Is it worth it? Are the inconveniences, extra time, and perhaps discomfort involved in taking all the safety precautions really worth the effort? Besides, taking the risk or shortcut is rewarded with soon and certain positive consequences, such as comfort, convenience, and efficiency. We get the job done faster and enjoy a longer break.

Take texting again. You're stuck in traffic or traveling a little above the speed limit on an open highway. In both cases you have windshield time—the perfect opportunity to send a text-message or make a phone call. The soon and certain positive consequences of this at-risk behavior take control. They overpower any thought that this risky behavior could cause a vehicle crash. And in most cases, a crash does not happen and the at-risk behavior is rewarded.

Thus, we may believe the numerous safe behaviors promoted by safety pros are good for us and an injury-free workplace requires consistent performance of these behaviors by *all* employees. Yet many workers still take risky shortcuts. Why? Because they believe it won't happen to them. They took the calculated risk many times before with no negative outcome, only positive consequences: comfort, convenience, and efficiency.

### The AC4P rationale

At this point, you might be saying, "I knew that," or "I've heard this before." In fact, if you heard or read an introductory presentation on BBS, you have certainly heard this logic before. This is the primary rationale for implementing a BBS process that targets the motivational aspects of at-risk behavior and attempts to hold people accountable for these behaviors through interpersonal feedback conversations and accountability displays. Let's address the nature of these conversations with regard to the three-question empowerment model.

You assess self-efficacy and response-efficacy with training and education, respectively. Indeed, a most frequent corrective-action plan after an injury investigation that reveals the occurrence of unsafe behavior is, "The employee will be retrained."

Of course, it's possible that the employee did not know the safe way to complete a job task, but it's more likely self-efficacy was not the issue, and neither was response-efficacy. The critical issue was outcome-expectancy, or motivation by expected behavioral consequences, as discussed above. The worker likely knew how to perform the task safely and actually believed the safe way was optimal for injury prevention, yet the worker was just more motivated to take the risk.

Let's return to our AC4P conversation. What rationale could you offer that would overcome this motivation to take the risky shortcut? The common reason we've heard, "You might get hurt," is not effective because the person's experience discredits this assertion. The person will say, "I've taken this risk many times before, and because I'm experienced and attentive to possible errors, I've never been hurt."

We suggest adding an AC4P dimension to this conversation. "Yes, I understand that your skills help you handle the risk of injury and thus it probably won't happen to you. But what if everyone performed the at-risk behavior? Would someone get hurt? Do you care?"

The employee will probably answer yes to the AC4P question. In fact, a "No, I don't care" reaction would indicate the kind of employee who is detrimental to achieving an AC4P safety culture and should probably receive a punitive consequence for the unsafe behavior observed. Indeed, you would hope employees with this mind-set would find another place to work.

But, we suspect and hope that you get a "Yes, I care" response. When you do, you can follow-up with reference to observational learning (Lesson 2). Indicate that the individual's skills and experience make him or her model employee whose behavioral example influences the action of others, especially less-skilled workers. When respected employees work at-risk, they help to establish a social norm to perform the risky behavior. The unsafe behavior is practiced plant-wide, and someone is eventually injured.

It's critical for employees—especially supervisors—to understand that their at-risk behavior sets an example for others to follow. Eventually, that risky behavior will contribute to an injury, and everyone who has practiced that behavior has contributed to that injury. So even if your at-risk behavior has never hurt you, it's possible it will injure someone else through observational learning and the establishment of an undesirable social norm.

Hence, the spirit of AC4P proclaims, "Yes, it's worth it. The inconvenience, inefficiency, and discomfort of a safe behavior are always worth it because my behavioral example influences the behavior of others, and I care."

**An emotional connection**

Okay, so how can we get people beyond their verbal affirmation? How can an AC4P emotion be activated? Note the term *emotion*. Feeling an emotion is motivational. We are motivated to enhance positive emotions and avoid or reduce negative emotions. In other words, we act to experience positive emotions and to remove the factors that activate negative emotions.

Many readers from the world of safety have heard Charlie Morecraft, Tony Crow, and Brad Gardner tell their heart-rending stories of their severe on-the-job injuries, each influenced by their own at-risk behavior. (If you haven't heard their stories, google them.) We have seen these individuals' moving presentations to large audiences on several occasions, and every time our emotions and those of the other spectators are dramatically aroused. Often there's not a dry eye in the house.

The emotional reactions of the audience verify that those folks care. Now is an opportune time to tell those emotional spectators how to *actively* care. What can they do to reduce the chance these extremely painful and debilitating injuries will not happen to anyone at their workplace? Now they have the motivation and are primed for self-efficacy. Their emotions communicate the why. Now they are ready to learn what to do in order to prevent the occurrence of similar injuries to themselves and others.

Here we started with answering the motivation question first (Is it worth it?) and now the audience wants self-efficacy. What can we do to prevent such workplace injuries? Many companies have hired these and other victims of severe injuries to tell their tragic stories, obviously to motivate workers to do more for safety at their organization. What else can companies do to arouse these "I care" emotions among employees?

Imagine a workplace where employees willingly talk about their own injury-related experiences, along with the environmental, management, behavior, and attitudinal factors that contributed to their close call or injury. It could have been a severe injury or even a fatality, but for some fortunate chance factors. Wouldn't this presentation be more motivating and instructive than an injury story from a speaker from another organization and work culture?

Here's the most important question: How can you develop the kind of work culture where employees feel comfortable telling and re-telling their injury or close-call stories, and feel appreciated when they do?

We sincerely hope you will learn practical answers to this question from this teaching/learning manual. In fact, we believe you already have answers, and we recommend you write these in the space below. It would be very useful to discuss your answers with those in your study group. This exercise could be invaluable in developing action plans at your facility to cultivate an AC4P safety culture and an injury-free workplace.

______________________________________________

______________________________________________

______________________________________________

______________________________________________

______________________________________________

## Empowering Goals

As an activator of process activities aimed at achieving a particular outcome, behavior-focused goal setting facilitates individual and group success. Perhaps you are aware of a popular acronym used to define the characteristics of an effective goal: SMART. There are actually a few variations of the words reflected by these acronym letters, with *M* representing Measurable or Motivational, and *T* referring to Timely or Trackable.

We propose the following acronym, with an added *S* at the end: SMARTS: Specific, Motivational, Attainable, Relevant, Trackable, Shared. We believe that the Shared characteristic is important to include because social support can increase commitment to work toward reaching a goal. Please note the connection between SMARTS goals and the empowerment model introduced earlier on page 97.

Specifically, SMARTS goals are empowering because they are *attainable* ("I can do it"), *motivational* ("it's worth it"), and *relevant* ("it will work"). This connection makes it clear that both empowerment and goal setting are similar behavioral antecedents, setting the stage for or activating certain behavior. Each of these activators of behavior refers to motivation as the anticipation of a desirable consequence or outcome. Plus, verbal behavior that contributes to goal setting or feelings of empowerment would be considered feedforward, as discussed earlier.

## Empowerment versus Self-Motivation

It's important to consider a critical distinction between these two person-states. Empowerment is a behavioral antecedent (i.e., feedforward), whereas self-motivation reflects the impact of consequences (i.e., feedback). In other words, feeling empowered means the individual is ready, or activated, to work toward achieving a given goal. On the other hand, a self-motivated person is anticipating or has received a consequence (e.g., feedback) that supports self-directed rather than other-directed behavior.

With this perspective, consequences that reflect personal choice, competence, and/or a sense of social support or community should enhance self-motivation and thereby increase the durability of a behavior-change intervention that applies positive consequences (Lesson 1). Thus, an intervention applying positive consequences to increase the occurrence of a safety-relevant behavior will have longer-term impact if the intervention inspires self-motivation by linking the behavioral consequences with a perception of choice, competence, and/or community.

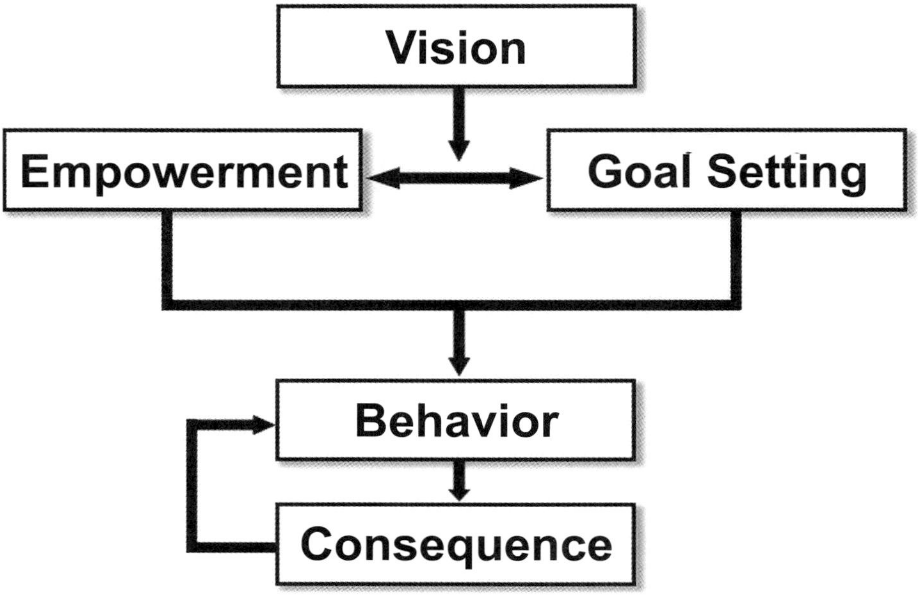


**Figure 7.** The connection between empowerment, goal setting, and consequemces

Figure 7 illustrates how empowerment, vision, and goal setting fit with the ABC (activator-behavior-consequence) model discussed earlier in Lesson 3. The simple but fundamental point: A vision and a goal are not sufficient to sustain desirable behavior. People need to feel empowered to work for goal achievement, including the acquisition of desirable consequences and/or the avoidance of undesirable consequences.

Participants need to believe in and own the vision. They need to feel support from peers to attain process goals that support the vision. And peers need to give them supportive and corrective feedback to increase the quantity and improve the quality of behaviors consistent with vision-relevant goals. Note that behavioral consequences are crucial. Empowerment and goal setting can activate the occurrence of desirable behavior, but without relevant supportive consequences, the behavior won't last. It will extinguish.

## Pause for a Review and Discussion

At this point, we encourage you to review the key components of this lesson (self-motivation, empowerment, and goal setting) by watching a 15-minute TEDx talk on YouTube. Simply google Scott Geller TEDx, or use the link: https://youtu.be/7sxpKhIbr0E.

Afterward, the instructor should facilitate a discussion of answers to the questions listed below. Participants should either write personal answers to the questions, or break up into smaller interactive groups and discuss answers among themselves before giving small-group reports to all participants.

Whatever format is most appropriate, the objective is to activate some interactive engagement whereby participants express their viewpoints and verbalize connections between the self-motivation concepts in the TEDx talk and their own life experiences.

## Discussion Questions

1. **Explain the three beliefs that determine empowerment as a person-state and connect each belief to ways for increasing empowerment through training, education, and motivation.**

2. **Explain the three perceptions (i.e., the three *C*s) that influence self-motivation and provide examples from personal experience.**

3. **Explain the meaning and significance of what B.F. Skinner referred to as "selection by consequences."**

4. **Distinguish between a *goal* and a *vision* with a personal example.**

5. **The ABC model of behavioral science presumes behavior is directed by activators (or feedforward) and motivated by consequences (feedback). How can goal setting provide both direction and motivation?**

6. **Explain the difference between independence and interdependence with regard to achieving and sustaining an injury-free work culture.**

**7. Which of the lessons in this brief talk on self-motivation were most meaningful to you? Please explain why?**

# Lesson 7

## PROGRESS FROM SELF-ACTUALIZATION TO SELF-TRANSCENDENCE

The most popular theory of human motivation is the hierarchy of needs proposed by humanist Abraham Maslow. Categories of needs are arranged hierarchically, and it is presumed people don't attempt to satisfy needs at one stage or level until the needs at the lower stages are satisfied.[63]

First, we are motivated to fulfill physiological needs. This includes food, water, shelter, and sleep for basic survival. After these needs are met, we are motivated by the desire to feel secure and safe from future dangers. When we prepare for future physiological needs, we are working proactively to satisfy our need for safety and security.

Next, we have our social-acceptance needs—the need to have friends and feel like we belong. When these needs are gratified, our concern shifts to self-esteem, the desire to develop self-respect, gain the approval of others, and achieve personal success. After that we strive for *self-actualization*. Is this the highest level of Maslow's hierarchy of needs? No, it is not.

### The Highest Need Level

When you look at the pyramid illustrating Maslow's hierarchy of needs (figure 8), you'll notice that self-actualization is not at the top. Maslow revised his renowned hierarchy shortly before his death in 1970, placing self-transcendence above self-actualization.[64] Transcending the self means going beyond self-interest to actively care for others.

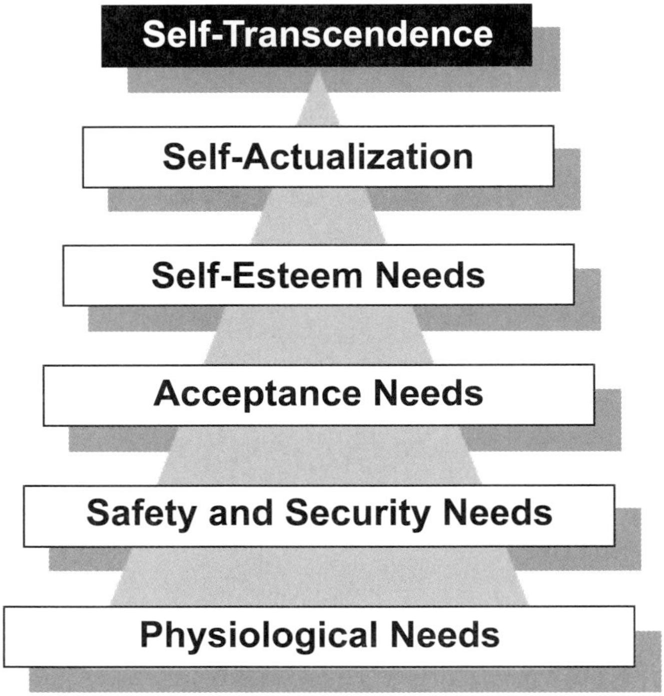


**Figure 8.** Maslow's revised hierarchy of needs, including self-transcendence

It seems intuitive that various self-needs require satisfaction before self-transcendent or AC4P behavior is likely to occur. But scant research supports the ranking of human needs in a hierarchy. It's possible to think of many examples of individuals performing various AC4P behaviors before satisfying all their personal needs. Mahatma Gandhi put the concerns of others before his own. He suffered imprisonment, extensive fasts, and eventually assassination in his fifty-year struggle to help his poor and downtrodden compatriots.

Note the connection between the need hierarchy and various rewarding consequences, as well as our discussion of self-motivation and sustaining the impact of a behavior-improvement intervention. One's position in the hierarchy certainly determines what types of consequences are likely to be most rewarding at a particular time. Without food, shelter, or sleep, for example, we would focus our efforts on satisfying these needs. But if this need-level is satisfied, the motivation of human behavior requires consequences related to higher-level needs.

As we climb Maslow's hierarchy, need states are reached that implicate consequences linked to self-motivation. For example, consequences that boost our sense of connection with others (i.e., community), satisfy the need for acceptance or social support, and consequences that certify our belief in personal competence to perform worthwhile work are associated with the self-esteem and self-actualization needs. Plus, it's intuitive that reaching beyond self-needs to help others through AC4P behavior can contribute to satisfying our need for social acceptance and self-esteem, and even self-actualization.

When does your need for social acceptance, self-esteem, and self-actualization get satisfied? In other words, at what point do you become satiated on consequences linked to these need states? Yes, these are rhetorical questions, posed to reiterate the value of delivering rewarding consequences that reflect the three *C*s of self-motivation (Competence, Choice, and Community) and thereby enhance need states in Maslow's hierarchy that are difficult to satiate.

Bottom line: Behavioral consequences that foster perceptions of personal competence, self-worth, interpersonal belongingness, and/or autonomy also facilitate self-motivation and self-directed behavior, and are thus likely to have a longer-term impact than consequences unrelated to these person-states.

## Discussion Exercise

1. **It would be helpful to discuss practical answers to the fundamental question implicated above: What kinds of behavioral consequences support or enhance your personal perceptions of choice, competence, and community with regard to occupational safety and health?**

**2. In many cases, the verbal delivery of supportive feedback (as discussed earlier) can influence the recipient's self-motivation. Please specify below the verbal support you give to others or receive from others that facilitate the three *C*s of self-motivation: Choice, Competence, and Community.**

# THE AC4P MOVEMENT

Most of us care about the well-being of others, but too few of us *act* proactively on these feelings of caring. How can everyone be activated and motivated to *actively care*, to take effective action based on our caring.

Our vision is to engender a brother's/sister's keeper culture in which we all look out for each other's safety, security, and well-being—a culture in which people routinely go above and beyond the call of duty for the health, safety, and/or welfare of others. This is an actively caring for people (AC4P) culture.

## The AC4P Wristband and SAPS

How can supervisors, safety professionals, and line workers nurture interpersonal empathy, compassion, and AC4P behavior? One way is by wearing the AC4P wristband, which signifies your support of the AC4P movement. When you reward the desirable behavior of another person with an AC4P wristband, you promote and fuel the AC4P movement.

When awarding an AC4P wristband, you should be very specific about defining the desirable safety-related behavior you observe and appreciate. Hand the wristband to the individual as recognition for setting an AC4P example. Give the wristband with words that connect to a higher-level need.

Don't suggest or even imply that the wristband is a payoff for AC4P behavior. That's never the reason. "The wristband is a token of appreciation for the special servant leadership exemplified by the act of kindness observed." Tell the person he or she is now one of many who have joined the AC4P movement, a flourishing worldwide effort to cultivate cultures of interpersonal compassion and interdependent AC4P behavior.

Proceed to explain the SAPS process—See, Act, Pass, Share—so the individual is motivated to continue to support the AC4P movement. First, See someone looking out for the safety or health of another person. Then, Act to recognize and reward this person for the safety-related behavior you observed with an AC4P wristband. Say you are passing on your AC4P wristband for the kind act you just witnessed.

Ask the individual to Pass the wristband to another person he or she observes performing a safety-related behavior consistent with the AC4P movement. Finally, please ask this wristband recipient and new member of the AC4P movement to Share this positive exchange on the Internet at www.ac4p.org. There the person can register the number of the wristband and briefly describe the safety-related behavior that led to the delivery of the AC4P wristband. When this wristband is passed on, a new AC4P story tied to that specific wristband number is posted on the AC4P website.

In this way, positive gossip is spread at your workplace and beyond. Others see that acting beyond one's self-serving interests on behalf of another person's safety is more common than imagined. AC4P storytelling contributes to making AC4P behavior a social norm. It creates and nurtures a culture of interpersonal trust, compassion, and routine AC4P behavior.

To be sure, the SAPS process is much easier said than done. Why? Actively caring behavior is not yet part of our normal routine, and it's not expected of us. We do not normally look for safety-related behavior to reinforce, even though this is the most powerful way to improve behavior (Lesson 1). In fact, we commonly deny even a thank you given us for a kind act with expressions such as, "No problem," "Don't worry about it," or "Just doing my job."

Receiving an AC4P wristband from a work colleague or supervisor will be very rewarding and memorable. The one-on-one conversation will be impactful, especially the AC4P message accompanying the wristband reward. Whenever we explain the AC4P concept to others, it's always appreciated. Whenever we offer an AC4P wristband, it's always accepted with a sincere smile and worn with a sense of pride. In fact, we know many individuals who wear their AC4P wristband every day and actually resist passing it on.

Imagine an individual receiving an AC4P wristband from a coworker who gives a gracious and tactful description of the AC4P behavior that justified this recognition. Such rare and unexpected recognition is sure to be accepted with pleasant surprise. In turn, this wristband recipient will tell friends and family about this very positive and unique encounter at work. More important, the one-on-one positive interaction will likely increase the frequency of the safety-supportive behavior that was recognized.

As these positive exchanges among employees accumulate, the media will take note and report the positive news about the AC4P approach to occupational safety. The AC4P movement then spreads throughout the community, and employees at your organization are identified as the agents of this research-based approach to promoting acts of kindness—actions that not only contribute to an injury-free workplace but also promote a culture with more interpersonal compassion and less interpersonal conflict, aggression, and violence.

## Discussion Questions

1. **What are some safety-related behaviors an employee might observe from others that warrant the reward of an AC4P wristband?**

2. **What would you say to the person when rewarding him or her with an AC4P wristband for performing a safety-supportive behavior?**

## Feedforward and the SAPS Process

The AC4P wristband and SAPS process applied to support occupational safety and health is essentially a feedback process to show appreciation for and reward AC4P behavior that contributes to the achievement and maintenance of an injury-free workplace. Supervisors and line workers search out AC4P behavior from other employees that support safety. When they observe this behavior, they immediately seize the opportunity to reward the AC4P behavior and solicit another participant for the AC4P movement.

But how often do you have an opportunity to actually observe safety-relevant behavior in situations where it's convenient or at least feasible to deliver an AC4P wristband and the AC4P movement message? Here's an alternative approach to distributing AC4P wristbands and promoting occupational safety and the AC4P movement. Deliver your AC4P wristband as feedforward rather than feedback. Consider giving AC4P wristbands to supervisors and line workers after *you* perform an AC4P behavior for them.

Specifically, after an employee thanks you for your AC4P behavior on behalf of his or her safety, security, or well-being, follow up by passing on an AC4P wristband to this person and asking him or her to join the AC4P movement. You might say something like the following:

> Thank you for appreciating the safety-supportive behavior performed by my work colleagues and me. Wouldn't it be nice if all of us performed more safety-related behavior on behalf of others? In fact, many employees at our plant have joined an actively caring for people initiative, or AC4P movement, to spread acts of kindness between people nationwide and, actually, worldwide. This wristband reflects this movement, and I hope you will wear it and join us.
>
> Every wristband is engraved with its own ID number. I will record the ID number on the wristband I've given you when I report this AC4P event at the AC4P website. Will you also report my safety-related act, along with the ID number on your wristband? The website address, www.ac4p.org, is engraved on the wristband.
>
> I hope you will look for opportunities to pass on your wristband, either after you actively care for another person's safety or when you see another person looking out for someone else's safety. In the first case, you acknowledge your own injury-prevention behavior, as I did with you. In the second scenario, you reward someone for his or her AC4P behavior on behalf of occupational safety and health.

Of course, the above is only a suggestion. You might certainly discover another way to express three key points: (1) explaining why you passed on the AC4P wristband to the beneficiary of your AC4P service; (2) introducing the AC4P movement and the reporting of AC4P stories at the AC4P website, along with the ID number on the wristband; and (3) encouraging this new member of the AC4P movement to pass on the wristband when he or she performs a safety-relevant act (as feedforward) or when he or she observes a safety-relevant AC4P behavior from another person (as feedback).

## Discussion Questions

After explaining this feedforward approach to passing on an AC4P wristband, discuss any perceived benefits of this process, other than having an opportunity to explain the AC4P movement and pass on an AC4P wristband.

1. **How can this feedforward practice generate more AC4P behavior for injury prevention? Jot down some possibilities here, and then discuss various answers among colleagues or workshop participants.**

2. **How would you feel if you were given the AC4P wristband? What would your immediate response be to an encounter like this?**

## Psychological Impact of Feedforward

Lessons 1 and 3 covered the rationale and benefits of rewarding people for their AC4P behavior. Simply put, supportive feedback can boost self-esteem, personal competence, and a sense of community or belongingness, and this positive exchange increases the likelihood that the individual will perform another safety-relevant act of kindness. Plus, you help that person bask in the reflected glory of reaching the highest level of Maslow's hierarchy of needs, self-transcendence (Lesson 7).

What about the feedforward method? What is the psychological impact of giving someone an AC4P wristband after you have helped that person? Let's consider the research-supported principle of reciprocity.[65]

Many people feel a need—even an obligation—to pass on a good turn after receiving one from another person, according to much research evidence provided by social psychologists.[65] When possible, the favor is returned to the original benefactor. But when this is impossible, as when a stranger contributes an AC4P behavior, the beneficiary of the kind act can satisfy the need to reciprocate by helping someone else, even a total stranger. In other words, if you are stuck in a traffic jam and a driver kindly allows you to pass in front of him you might then show that same kindness to someone else, perhaps another stranger stuck in the same traffic jam as you.

These types of acts of kindness can spread, leaving positive feelings and encouraging a ripple effect of kind acts. Thus, the reciprocity norm and related research suggest the AC4P feedforward wristband influences the recipient to perform an AC4P act.

The principle of consistency is also relevant here.[65] People want to be consistent in their actions and attitudes. This research-supported principle indicates it's possible to *act* a person into a certain attitude (or way of thinking), and vice versa. When people perform an act of kindness and/or pass on an AC4P wristband, their positive thoughts and attitudes about the AC4P movement get a boost.

Every time a supervisor or line worker distributes an AC4P wristband as feedforward or as feedback, a deposit is made in the recipient's attitudinal or emotional bank account for the AC4P movement. That person will then want to behave in a manner that is consistent with how others perceive them. In this case, the recipient becomes more committed to the AC4P mission of building a brother's/sister's keeper work culture and sustaining an injury-free workplace.

How can the principle of consistency influence AC4P behaviors for safety and improve attitudes of someone who receives a feedforward wristband? You know the answer, right? By accepting the wristband and putting it on, the individual performs behaviors that reflect a positive attitude toward the AC4P movement. If this person later logs on to the AC4P website and reports the positive exchange, more behavioral deposits are made to support consistent and positive self-talk, attitudes, and affect toward occupational safety and the AC4P movement.

Plus, this individual might receive supportive feedback regarding this website posting, since stories posted at the AC4P website are forwarded to that individual's Facebook page. Such social support could serve as more feedforward to influence more AC4P behavior for safety and other domains in need of more interpersonal acts of caring.

Eventually the continual spiraling of interpersonal feedforward fueling AC4P behavior that is supported by interpersonal supportive feedback can result in genuine personal commitment to the AC4P movement and the achievement of a brother's/sister's keeper culture at work, at school, at home, and throughout the community at large.

## Discussion Questions

The instructor should facilitate interactive discussions of the following questions, perhaps in small groups and followed by reports to the entire audience.

1. **What positive short-term and long-term consequences are likely if the managers and line workers of an organization successfully spread the AC4P movement throughout their workplace and beyond? Note how answers to this question can motivate individual and group execution of AC4P behavior (Lesson 1).**

2. **What is the value of documenting each delivery and receipt of an AC4P wristband on a website?**

3. **How could employees be motivated to record their SAPS experiences at the AC4P website?**

4. **What factors might hold someone back from initiating and maintaining the SAPS process? How can these potential obstacles be minimized?**

5. **What kind of support could facilitate the large-term success of an AC4P movement for safety in your organization?**

6. Is there more to the AC4P movement than the SAPS process? Please explain.

7. What are the most important takeaways from this education/training program? In other words, what principles and/or procedures covered in this handbook and discussed in your education/training program are most useful to you?

8. In what ways are the principles applied in the AC4P movement useful beyond your job? How might you apply these principles or procedures in situations beyond your organization?

**9. What did you like best about this education/training process?**

______________________________________________

______________________________________________

______________________________________________

______________________________________________

______________________________________________

**10. How can this education/training for cultivating an AC4P brother's/ sister's keeper culture be improved?**

______________________________________________

______________________________________________

______________________________________________

______________________________________________

______________________________________________

## Role-Playing Exercise

This final role-playing exercise is most important. Participants should practice the interpersonal communication steps of delivering an AC4P wristband as both feedforward and feedback, and explaining the SAPS process.

This can be accomplished by having participants pair off and play the roles of giver and receiver of a feedforward and feedback delivery of an AC4P wristband for performance of a safety-relevant behavior. Later, volunteers can replay their interactions before the entire group of participants, followed by feedback from the instructor and other observers.

Or, participants can stay with their discussion groups and develop a role-play demonstration for the entire group. It would be enjoyable and perhaps instructive to create a role-play of both wrong and right ways to deliver and receive an AC4P wristband as feedforward and feedback.

The purpose: It's never easy to interrupt an individual working or in the middle of their day, so stopping them with a message can take a lot of courage and practice. This kind of role play can help participants feel comfortable giving people an AC4P wristband as feedforward and feedback and explaining the SAPS process.

As mentioned, achieving this level of comfort is challenging, but participation should be motivated by realizing the positive consequences from this AC4P process: increasing the occurrence of safety-relevant behavior at your workplace and promoting and supporting an AC4P movement designed to cultivate cultures of compassion and mindful acts of kindness worldwide.

# CONCLUSION

Congratulations! You have just learned leading-edge principles and procedures for improving behavior relevant to keeping people safe, while also enhancing positive connections between you and other employees at your work facility.

We sincerely hope you have acquired more than an understanding of the seven principles of humanistic behaviorism (the academic term for the foundation of the AC4P movement). We hope you *believe* in the validity of these research-based principles to improve interpersonal attitudes and behavior related to increasing the frequency and improving the quality of safe behavior, decreasing the frequency of at-risk behavior, and cultivating an AC4P injury-free work culture.

Most important, we hope you feel *empowered* to begin practicing the principles with your family, work colleagues, and eventually with perfect strangers you observe beyond the workplace. Implementing the feedforward and feedback techniques with empathy will surely reap observable benefits. Plus, when you reflect on the results of your behavior-focused conversations, you continuously improve your skills at using one-on-one conversation to benefit the behavior of others.

When you add the AC4P wristband and the SAPS process to your safety-related communications with colleagues at your work site, you enhance the positive consequences of each conversation. You will have recruited another participant for the AC4P movement and helped to nurture an AC4P culture of interpersonal trust, empathy, compassion, and routine acts of kindness.

The result will be a work culture in which everyone steps up to the plate on behalf of the safety and wellness of others. Coworkers will offer supportive feedback for safe behavior and deliver corrective feedback for at-risk behavior with empathy.

Everyone accepts and appreciates feedforward reminders to perform a certain task safely. Coworkers activate the *I care* emotions of other coworkers when they discuss their close calls and minor injuries that could have been much worse and point out situational and dispositional factors that contributed to the mishap.

Behavior is involved in almost every workplace injury and needs to be addressed daily in order to sustain an injury-free workplace.

Brother's/sister's keepers realize that everyone unknowingly performs at-risk behavior at times and needs AC4P feedforward and feedback to remain injury free. It's just the behavior that needs some adjustment, not the attitude.

The attitude is actively caring for people through routine behavior-focused AC4P coaching that supports safe behavior, corrects at-risk behavior, and inspires these answers to the three empowerment questions: "Yes, we can do it," "Yes, it will work," and "Yes, it is definitely worth the effort."

# AC4P WRISTBAND STORIES

Actively caring for people (AC4P) is epitomized by a get-well card sent to a friend, even a stranger; a cup of tea and a listening ear; returning found merchandise to its rightful owner; or shoveling a pile of snow from a senior citizen's driveway. AC4P means we appreciate people for the acts of kindness they perform for others, either proactively to prevent possible harm or reactively after harm is experienced. When stressors and life's conflicts get us down, we need to know someone cares about our difficulties and wants to lift us up. And we need to reciprocate, and do the same for others.

AC4P makes a lasting difference in our lives. It's not about being judgmental, slashing people's self-esteem, and picking over faults. Not at all. AC4P values our uniqueness. It recognizes our human potential and plays positively to our strengths.

Empathy is a prerequisite of AC4P behavior. We need to perceive and care about the concerns or predicaments of others. When called on to care for others, we draw on our inner compassion and courage to act.

The following personal stories demonstrate principles and applications of AC4P. They will inspire you to carve out more personal time to actively care for your family, friends, colleagues, and even strangers. Each story illustrates how an AC4P mind-set brings out the best in both the giver and the receiver.

These testimonies bring home an important point: AC4P behavior strengthens your compassion and caring—you are the person giving. Imagine the AC4P cascade effect when ever-increasing numbers of individuals experience the rewarding, positive, and healthy consequences of actively caring for others.

Leaping to the self-transcendent state that sits atop Maslow's hierarchy of needs fulfills our lower-level needs, especially our need for self-esteem and belongingness. Self-efficacy, personal control, and optimism are also strengthened through AC4P behavior. Empowerment and self-motivation are fueled by these person-states, and vice versa.

Our greatest challenge is to help others truly feel the rewarding power of AC4P behavior. To do this, we must develop the context and implement contingencies to motivate initial stirrings of AC4P behavior. The positive consequences of the AC4P behavior witnessed and experienced time and again in the following personal stories support the premise that AC4P behavior can come naturally—it is organically rewarding and self-sustaining when nurtured within a supportive culture.

These personal stories are not due to the competence, commitment, and courage of a single heroic person. Friends, family, teachers, coworkers, and even strangers provide direction, mentorship, or supportive consequences. Each story illustrates how AC4P principles enable the development of an AC4P culture—a culture of compassion at work, at home, at school, and throughout our communities.

## Aly Neel's Metro Story

After living in D.C. for some time and riding the Metro during many rush-hour mornings and nights, I have become well aware of the unwritten rule, "You just ride."

On my way home from work one day, I catch the red-line train toward Union Station as usual. A young man, wearing a suit, sits inches away from me. He's so close I can almost touch him. I look up and notice he seems very upset—wringing his hands, shaking his head. Unintentionally, I stare at him. I try but can't look away; he seems on the verge of tears.

I immediately think, "What can I do?" I know I have to say something, but I'm uncertain how to reach out. We finally make eye contact, and I give him a smile—the empathic kind I would give a friend whose family member just died. I want him to know I'm sorry for whatever he is going through. Immediately after our exchange I look down, sort of embarrassed. I remember people aren't *supposed* to smile at each other on the Metro!

The Metro slows down and comes to a stop. The guy, still shaken up, stands to get off the train, but then pauses to touch me on my shoulder. He says: "You probably already forgot what you did. It didn't seem like a big deal, but this year has been the worst year of my life. What you just did a second ago, though really small, is probably the most anyone has reached out to me in this past year."

Rolling up his shirtsleeve, he tells me, "It represents a pay-it-forward notion." He hands me a green wristband engraved with the words "Actively

Caring for People." My mouth is agape. I had heard of this AC4P movement, but I had never received a wristband until now.

## AC4P Internet Stories

Stories posted at ac4p.org tell of exchanges between people receiving and giving AC4P wristbands after specific AC4P behaviors. Wristband recognition occurs for simple gestures—such as holding the door for a stranger—as well as more complex acts requiring skills, financial stability, and time. These acts occur in many locations, including schools, restaurants, highways, community streets, and stores.

Many stories reflect competence, commitment, and courage. They involve ordinary people acting for the benefit of family, friends, coworkers, and strangers in both reactive circumstances and proactive situations.

Some individuals receive a wristband for AC4P in reactive situations, such as standing up for a friend after hearing a racist remark (Wristband #407); rebuilding a home after tornadoes devastated Joplin, Missouri (#4766); helping after a sibling's car broke down (#2974); and saving the life of a motorcyclist after a crash (#240).

Others are recognized for proactive AC4P behavior, such as helping a coworker wipe up a spill, reminding the passenger of an airport shuttle to buckle up, holding the door for a line of people entering a building, walking an intoxicated stranger to her home safely, refusing to continue a phone conversation after learning the other person is using a mobile phone while driving a vehicle, showing a coworker the safest way to perform a certain task, loaning a pair of safety glasses to a neighbor planning to split some logs, and offering a safety suggestion at a team meeting.

### Caring for victims of a vehicle crash

#### Wristband #21761

This woman stopped at the scene of a vehicle collision to see if everyone was okay. As a nurse, she checked the vitals of everyone involved and made sure that everyone was safe until the EMS arrived. She went out of her way to help those in need.

#### Wristband #125626

I am a police officer with the City of Norfolk. A vehicle crash occurred in the 300 block of E. Little Creek Drive. When I arrived on the scene, Ms. Janice Riddles was already on the scene tending to the driver, who was injured. She stayed with the injured driver, holding her head still and preventing further injury, until medics arrived and transported the driver to the hospital.

Ms. Riddles stated she is a nurse and was on her way home when she saw the disabled vehicle stopped in the roadway. She stopped her vehicle and immediately rendered aid to the driver. Ms. Riddles was very helpful to the police and medics by providing important information that resulted in assisting the driver of the crash.

### Wristbands #122169 and 122170

Heading home from the NAU HLC gym, my brother and I were quickly stopped by traffic. Upon noticing the wreckage up ahead, I quickly pulled our vehicle over and ran to the fallen motorcycle rider. A small crowd stood around the man, one person calling 911 and another kneeling before him assumedly checking his vital signs.

As I stood there, my initial thought was that he was already gone. I could not see life in his blank stare until, finally, one slow blink. Thankfully he was wearing a helmet (not required in Arizona)! After assessing the situation, confirming the man was tended to by someone trained, and ensuring help was on the way, I returned down S. San Francisco Street and began instructing the gridlocked drivers to turn around. My brother and I then directed traffic at the S. San Francisco/E. Franklin Avenue intersection for the following twenty-five or thirty minutes until police came and took over.

During this time, a man drove up in a white crossover, flashed his badge, and asked what we were doing. I was wearing sweatpants and a gym shirt while blocking a very main roadway. After quickly telling him what happened, he pulled over and informed my brother and me that he was the city of Flagstaff Chief of Police, and he gave us each an AC4P wristband and thanked us for helping.

### Wristband #12548

I arrived on the scene of a head-on car crash that happened directly in front of our vehicle. Several of my friends and I got out to help. I rushed to one of the cars that was seriously damaged and found a young girl inside. Both drivers' side doors were jammed shut.

When I went around to the passenger side, I discovered her legs were pinned between the seat and steering wheel/dashboard. While another passerby dialed 911, I got her to give me her parents' phone number. I called them to let them know what was going on.

I stayed with her and tried to keep her calm until the paramedics arrived. Later, after some time went by, she got my number from her parents and called. She said she had a wristband to give me and wanted my address. A short time later, I got the wristband in the mail.

**Wristband #99744**

One of my employees, Brian Day, was driving from Connecticut to Massachusetts on Route 84. He witnessed the outcome of two cars colliding, one of the cars flipping over a snow bank. Brian is a veteran fire fighter, and when he saw what happened, he sprang into action.

He approached the car and saw a woman trapped in a smoking vehicle that was on its side. He grabbed a metal bar from his truck, smashed the sunroof, and pulled the victim to safety. He then waited for emergency vehicles to get to the scene. He said that while he was there not one other driver pulled over to see if anyone needed help. Good work, Brian!

**Wristband #67255**

I attended a big gathering of high school link crew leaders today, and I was inspired by the story of a kid name AJ. He was at home and heard glass shatter in the road, and he checked to see what it was. He noticed it was a car crash. A girl was pinned underneath the car, and he pushed the car off her to save her life. His act of courage to help someone in dire need truly inspired me to be a better person, and I really look up to him for what he did. I thought he was the perfect person to pass on my wristband to.

**Wristband #240**

I witnessed a man wreck a dirt bike, causing him to crash through a glass window. Once I heard the breaking glass, I ran to the scene and saw lacerations on his arm and several on his leg. I knew this was serious when I saw the amount of blood he lost in the 20 seconds or so it took for me to get there.

I, along with another Appalachian State University student, used t-shirts to stop the bleeding and make him comfortable until the paramedics arrived. He received more than 200 stitches for all his wounds. We were told he would have bled out if we had not stopped the bleeding right away.

**Wristband #17630**

It was the day after Christmas, at 6:00 on a Sunday morning. I'm a nurse and was driving into Baltimore for work. I was cruising on I-95 just like any other morning and saw a car that appeared stalled in the middle of the interstate about 100 feet ahead. I pulled into the right lane and slowed down. As I approached, I realized the car was totaled and none of its lights were on.

I immediately pulled onto the side of the road and reached into my pocket to pull out my phone to dial 911. Before I was able to call, a woman squeezed out of the wrecked car and came running to the side of the road where I was, holding her chest. I got out of my car and asked what had happened. She told me her car was hit by another car, causing her to spin, and then a tractor-trailer slammed into her vehicle. Both the other car and the tractor-trailer drove off, leaving her car smashed in the middle of the road.

I got on the phone with the police while keeping her calm and assessing her to make sure she was all right. As this was going on, cars were weaving around her car, which was still in the middle of the road. Suddenly, an SUV crashed into her car, causing it to go flying to the side of the road about 30 feet from where we were standing. That's when I truly realized the danger of the situation.

The SUV driver got out and came running to where we were. I helped keep both of them calm, got their medical histories, and assessed them for injuries—all before the police and an ambulance finally arrived. As the woman was being loaded into the ambulance, I noticed she too was wearing scrubs and was a nurse on her way to work, just like me.

I told my brother this story today, and when I did, he pulled off his wristband and gave it to me, telling me to tell my story and pass on the wristband. So that's what I will do.

## Proactive AC4P behavior at work

**Wristband #54736**

Jair Flores was given this wristband by Warren Hubler on December 18, 2014, for asking a coworker to put on his safety belt while driving between work locations in September of 2014 in Montana. They were involved in a car crash ten minutes later. The safety belt saved the life of his coworker! Well done, Jair!

**Wristband #56182**

As a contractor on the floor of the factory I work at, I witnessed an employee demonstrate AC4P behavior in the context of ladder safety. There was an employee working nearby who was seated on a ladder and working on an electrical junction box. There was nobody assigned to spot this worker, and he had positioned the ladder perpendicular to his work. (If he had pushed against the work, the ladder would have easily tilted over.) The Actively Caring employee quickly identified the situation and took on the personal responsibility to advise the other employee of his at-risk work practices.

**Wristband #56152**

While visiting a factory as an outside vendor, I was almost run over by a forklift. Fortunately, one of the very observant workers was able to quickly identify the hazard and pull me out of the way. Thanks to his quick action I was able to return home safely to my family and work the next day. Thank you!

### Helping a classmate combat a bully

**Wristband #19151**

Jerlany, a student at Fairchild Wheeler Multi Magnet High School, noticed a student being bullied while the teacher was in the hallway dealing with a disruptive student. She repeatedly asked the bully to stop and to leave the other student alone as the rest of the class either laughed along with the bully or became bystanders. When the bully pulled the student's chair out from under her, Jerlany came to the student's aid and escorted her to the bathroom and gave her words of encouragement as she recomposed herself after becoming visibly upset. I gave this wristband to Jerlany for her courage to help a student in need when no one else seemed to care.

### Saving a friend from a negative situation

**Wristband #75854**

Around 4:00 A.M. my friend texted me, asking me to call the police for her. I bake very early on Saturdays for my business, so she must have figured I'd be awake. I was glad I was. Her roommate and his girlfriend had been fighting, and the situation was turning violent. She was scared and locked in her room.

I phoned the police and later stopped production to pick her, her dog, and her other roommate up to sleep at my apartment. They got settled in, and I had to go back to work. I wish I could've stayed longer but was just glad to know she was safe. She received this wristband at a memorial/vigil for a friend, and I'm really touched that she passed it on to me. I'm excited to keep passing it on.

### Watching a friend's back

#### Wristband #94

Before I get to the AC4P moment, it's worth mentioning that all these guys I was with—Zach, Amin, Rob, and Henry—are not only some of the coolest and funniest guys I know, but they are true gentlemen to me. We've been friends since high school, and they are always looking out for me. They take care of me.

The story: My friends and I were visiting Zach and Amin in Boston. It was my first time there, and they showed us all the sites, including Fenway Park. When we were walking around Boston, I was sort of the typical tourist/runt of the group, falling behind and having to catch up all the time because of all the pictures I wanted to take. Every block or so one of the guys would look back, make sure I was okay, and wait up for me; I gave Henry a wristband for that the first day we were there.

So, it was our last day in Boston, and Zach/Amin took us to Fenway Park. I was right underneath the sign, and I lifted my camera to my face to take a picture. I was focusing in on the image when, all of a sudden, Zach yelled my name so loud it frightened me. I immediately turned to look and coming at me was a semi-truck. The driver didn't see me because he was backing up, and I got out of the way *just* in time.

If Zach hadn't looked back to see if I was safe, I probably wouldn't be here. I really think it says something about his character, and the character of our group of friends. Zach intentionally actively cares for the people around him.

### Avoiding the bystander effect

#### Wristband #77634

I was waiting outside Cramwell International Center for DMV services when I noticed something quite odd. A man was sitting with his head down and his shoulders hunched over. His passport was on the ground, and he seemed to be completely out of it. Something did not seem right, yet no one noticed this odd behavior. He was in clear view, everyone could see him, and there were about twenty people there. Yet no one got up to help or ask if something was wrong. I went over and asked him if he was okay. He did not reply. I asked again. No reply. I shook him lightly and he just collapsed on the bench. I yelled at someone across from me waiting for the DMV services to call 911.

Only two people came over to help, and the rest just stared. That really frustrated me because they seemed more worried about losing their place in line than about the health of another human being! It was shocking. When the police and the ambulance came, I explained how I found him. He began to regain consciousness, but he could not get up without falling back down. It seemed like a case of dehydration or malnutrition. Either way, he ended up being taken care of. But the question is: Who would've stepped in if I weren't there?

### Promoting positive coping

#### Wristband #57686

I am John Eklund, the Ohio State Senator representing much of the Chardon School District, where tragedy struck in February 2012. I received my wristband from a Chardon High School student after I made some remarks and presented an Ohio Senate Proclamation at a Prevention Network Rally on the steps of the Ohio Capitol on May 2, 2013.

My remarks and the Proclamation were meant to honor the Chardon students and others attending for the way they face adversity and personal challenges by coming together and supporting each other, rather than wallowing in their own self-pity or turning to drugs to cope. In that sense, I was only expressing what so many of the people in our Chardon community feel, believe, and live—my actions were just a recognition, not anything heroic or special. To think that they moved someone to honor me with a wristband is very special, but wholly undeserved. Those young people are the ones to be honored, as I will honor someone when I pass on the AC4P wristband.

### Compassion during times of hardship

#### Wristband #15260

My sister's husband recently passed away after an extended illness. As she completed the difficult task of going through his closet, she wondered what do to with his nearly new (and even some brand new) business and casual clothes.

She noticed that the bus driver for the faith-based school where she teaches always wears t-shirts and jeans. She was aware this was because of financial hardship and was not a fashion statement. She asked his size and he wears the same size as her husband's new clothes.

She gave him all the clothes that fit him—outfitting him with an entirely new wardrobe. He and his family are so grateful. I've sent the wristband to my sister, and I know she'll not only pass it on, but continue (as she has) to actively care for people. She is an inspiration to me and all our family.

### Strengthening friendship

#### Wristband #12576

I recently passed a wristband on to one of my friends at Summer Residential Governor's School. A girl I know was giving me a hard time while a bunch of our friends were spending time together. I left the room, clearly upset, to spend the rest of the night in my dorm rather than provoking the girl even more. My friend, agreeing the girl's comments were out of line, came up to spend the rest of the night in my dorm room with me. We talked mostly about other things.

I thought it was incredibly sweet of my friend to go out of her way to cheer me up. By giving her the wristband, I let her know she means a lot to me as a friend and I really appreciate her. She got a little teary-eyed (in happiness, of course). My first experience giving someone an AC4P wristband was one I will truly never forget, because I grew much closer with that friend as a result.

### What goes around comes around

#### Wristband #10534

I have always been a proponent of *pay it forward*, and when I initially heard of AC4P, I felt like a younger generation now had the chance to pay it forward and understand the benefits.

My story began last winter at a restaurant. I saw a family who was told by the hostess to stand outside in the cold. I quickly finished my dinner and asked that they be seated at our table. The family was from out of town, visiting their daughter, and was truly appreciative.

Fast forward to this summer: My youngest daughter and I were at a local supermarket. I had forgotten my wallet but had checks. However, since I did not have my license, the clerk told me to put my groceries back. Out from behind in the line, a young female said, "I will pay for her." I thanked her and tried to write her a check, but she would not take it.

Then, a different young female came from the door and ran up to the girl who paid for me. She said, "You are actively caring, here's a wristband for you, pass it on." I screamed with joy, and my seven-year-old child smiled.

After leaving, I realized the girl who paid for me in the grocery store was Catherine—the same young girl to whom I had given my seat at the restaurant six months earlier. Pay it forward! Actively Caring for People can become a global movement with your help!

### From one AC4P act to another

#### Wristband #57071

A few weeks ago, Dr. Geller came through the door, buzzing with excitement. He had just been to his bank, and Dalton, one of the bank tellers, told him she had a story she needed to share.

She had been at Panera Bread, venting on her cell-phone about an unpleasant event. Frustrated, Dalton hung up after saying, "I'm in Panera now. I'll call you later." After she ordered, the man behind the counter smiled and told her he'd pay for her meal because she was having a bad day. Grateful, and her day brightened, Dalton prepared to pass on the wristband she'd received weeks before only to realize she'd left it at home.

She explained this to him, saying she wanted to pass the AC4P wristband on to him. Matt proudly displayed his wrist, revealing a green AC4P wristband, "It's all right. I already have one." He had received *Wristband #57071* two weeks earlier from Joanne Dean Geller.

# REFERENCE NOTES


1. E. S. Geller, *The Psychology of Safety: How to Improve Behaviors and Attitudes on the Job* (Radnor, PA: Chilton Book Company, 1996); E. S. Geller, *Applications of Behavior Analysis to Prevent Injury from Vehicle Crashes*, 2nd ed., monograph published by the Cambridge Center for Behavioral Studies (Cambridge, MA, 1998); E. S. Geller, "Behavior-Based Safety in Industry: Realizing the Large-Scale Potential of Psychology to Promote Human Welfare," *Applied & Preventive Psychology* 10(2001):87–105; ed. E. S. Geller and J. Williams, *Keys to Behavior-Based Safety from Safety Performance Solutions* (Rockville, MD: Government Institutes, 2001); T. R. Krause, J. H. Hidley, and S. J. Hodson. *The Behavior-Based Safety Process: Managing Improvement for an Injury-Free Culture*, 2nd ed. (New York: Van Nostrand Reinhold, 1996); T. E. McSween, *The Values-Based Safety Process: Improving Your Safety Culture with a Behavioral Approach* (New York: Van Nostrand Reinhold, 1995); D. Petersen, *Safe Behavior Reinforcement* (Goshen, NY: Aloray, Inc., 1989); B. Sulzer-Azaroff, *Who Killed My Daddy? A Behavioral Safety Fable* (Cambridge, MA: Cambridge Center for Behavioral Studies, 1998); J. H. Williams, *Keeping People Safe: The Human Dynamics of Injury Prevention* (Lanham, MD: The Scarecrow Press, Inc., 2010).
2. A. M. Al-Hemoud and M. M. Al-Asfoor, "A Behavior Based Safety Approach at a Kuwait Research Institution," *Journal of Safety Research* 37(2006):201–206; R. M. Choudhry, "Behavior-Based Safety on Construction Sites: A Case Study," *Accident Analysis & Prevention* 70(2014):14–23; D. Chen and H. Tian. "Behavior-Based Safety for Accidents Prevention and Positive Study in China Construction Project," *Procedia Engineering* 43(2012):528–534; P. J. Foster, A. Parand, and J. Bennett. "Improving the Safety Performance of the UK Quarrying Industry Through a Behavioural-Based Safety Intervention," *The Southern African Institute of Mining and Metallurgy - Surface Mining*, (2008):265–282; A. C. Grindle, A. M. Dickinson, and W. Boettcher. "Behavioral Safety Research in Manufacturing Settings: A Review of the Literature," *Journal of Organizational Behavior Management* (2000), 20(1):29–68; F. Ismail et al., "Behaviour-Based Approach for Quality and Safety Environment Improvement: Malaysian Experience in the Oil and Gas Industry," *Procedia - Social and Behavioral Sciences* 35(2012):586–594; H. Lingard and S. Rowlinson. "Behavior-Based Safety Management in Hong-Kong's Construction Industry," *Journal of Safety Research* (1997), 2(4):243–256;

R. B. McAfee and A. R. Winn, "The Use of Incentives/Feedback to Enhance Workplace Safety: A Critique of the Literature," *Journal of Safety Research* (1989), 20(1):7–19; A. Parand and P. J. Foster, "Behavioural-Based Safety in the Minerals Industry: A Research Based Methodology Carried Out in the UK Quarrying Sector" (presented at Queensland Resources Council Mining Health & Safety Conference, Townsville, Queensland, 2006); D. Petersen, *Safe Behavior Reinforcement* (Goshen, NY: Aloray, Inc., 1989); B. Sulzer-Azaroff and J. Austin, "Does BBS Work? Behavior-Based Safety and Injury Reduction: A Survey of the Evidence," *Professional Safety* (2000), 45(7):19–24.

3. "The Sting in Blacksburg," *Roanoke Times & World News* (Roanoke, VA), April 19, 1995.
4. E. S. Geller, "If Only More Would Actively Care," *Journal of Applied Behavior Analysis* 24(1991):601.
5. E. S. Geller, "Actively Caring for Occupational Safety: Extending the Performance Management Paradigm," ed. C. M. Johnson, W. K. Redmon, and T. C. Mawhinney, *Organizational Performance: Behavior Analysis and Management* (Binghamton, NY: The Haworth Press, Inc., 2001), pp. 303–326; E. S. Geller, *The Psychology of Safety Handbook* (Boca Raton, FL: CRC Press, 2001); E. S. Geller, D. S. Roberts, and M. R. Gilmore, "Predicting Propensity to Actively Care for Occupational Safety," *Journal of Safety Research* 27(1996):1–8.
6. E. S. Geller and B. C. Kipper. *Actively Caring for People Policing: Building Positive Police/Citizen Relations* (New York: Morgan James Publishers, 2017).
7. E. S. Geller, *Actively Caring for People in Schools: How to Make It Happen* (New York: Morgan James Publishers, 2017); S. McCarty et al., "Actively Caring to Prevent Bullying in an Elementary School: Prompting and Rewarding Prosocial Behavior," *Journal of Prevention & Intervention in the Community* (2016), 44(3):164–176.
8. T. Ludwig et al., "Actively Caring for Safety in South Africa: The Power of the AC4P Wristband," ed. E. S. Geller, *Actively Caring for People: Cultivating a Culture of Compassion*, 5th ed. (Newport, VA: Make-A-Difference LLC, 2016), pp. 209–215.
9. E. S. Geller, M. J. Kalsher, and G. Lehman, "Promoting Safety Belt Use on a University Campus: An Integration of Commitment and Incentive Strategies," *Journal of Applied Social Psychology* 19(1989):3–19; J. R. Rudd and E. S. Geller, "A University-Based Incentive Program to Increase Safety-Belt Use: Toward Cost-Effective Institutionalization," *Journal of Applied Behavior Analysis* 18(1985): 215–226.
10. E. S. Geller, *Corporate Incentives for Promoting Safety-Belt Use* (Washington, D.C.: U.S. Department of Transportation, 1982); E. S. Geller, G. R. Lehman, and M. J. Kalsher, *Behavior Analysis Training for Occupational Safety* (Detroit, MI: Ford Motor Company, 1987).

11. D. Gray, personal communication to E. S. Geller, February 7, 1984.
12. P. J. Foster, A. Parand, and J. Bennett, "Improving the Safety Performance of the UK Quarrying Industry Through a Behavioural-Based Safety Intervention," *The Southern African Institute of Mining and Metallurgy - Surface Mining* (2008):265–282; A. Parand and P. J. Foster. "Behavioural-Based Safety in the Minerals Industry: A Research Based Methodology Carried Out in the UK Quarrying Sector" (presented at Queensland Resources Council Mining Health & Safety Conference, Townsville, Queensland, 2006).
13. A. M. Al-Hemoud and M. M. Al-Asfoor, "A Behavior-Based Safety Approach at a Kuwait Research Institution," *Journal of Safety Research* 37(2006):201–206; F. Ismail et al., "Behaviour-Based Approach for Quality and Safety Environment Improvement: Malaysian Experience in the Oil and Gas Industry," *Procedia - Social and Behavioral Sciences* 35(2012):586–594.
14. D. Chen and H. Tian, "Behavior-Based Safety for Accident Prevention and Positive Study in China Construction Project," *Procedia Engineering* 43(2012): 528–534.
15. R. M. Choudhry, "Behavior-Based Safety on Construction Sites: A Case Study," *Accident Analysis & Prevention* 70(2014):14–23; H. Lingard and S. Rowlinson, "Behavior-Based Safety Management in Hong-Kong's Construction Industry," *Journal of Safety Research* (1997), 2(4):243–256.
16. E. S. Geller, *The Psychology of Safety: How to Improve Behaviors and Attitudes on the Job* (Radnor, PA: Chilton Book Company, 1996); E. S. Geller, *The Psychology of Safety Handbook* (Boca Raton, FL: CRC Press, 2001); E. S. Geller, "The Psychology of AC4P Behavior," ed. E. S. Geller, *Applied Psychology: Actively Caring for People* (New York: Cambridge University Press, 2016), pp. 45–82; J. H. Williams, *Keeping People Safe: The Human Dynamics of Injury Prevention* (Lanham, MD: The Scarecrow Press, Inc., 2010).
17. E. S. Geller, "The Truth About Safety Incentives," *Professional Safety* (1996), 41(10):34–39; E. S. Geller, S. Perdue, and A. French, "Behavior-Based Safety Coaching: Ten Guidelines for Successful Application," *Professional Safety* (2004), 49(7):42–49; D. S. Roberts, "Actively Caring for Occupational Safety: Preventing Injuries with People-Based Safety," ed. E. S. Geller, *Actively Caring for People: Cultivating a Culture of Compassion*, 5th ed. (Newport, VA: Make-A-Difference, LLC, 2016), pp. 181–197.
18. E. S. Geller and R. A. Veazie, "Behavior-Based Safety versus Actively Caring: From Other-Directed Compliance to Self-Directed Commitment," *Professional Safety* (2014), 59(10):44–50.
19. E. Williams et al., "Maintaining Momentum: Increasing Employee Participation, Strengthening Leadership and Supervisor Support: Utilizing Behavior-Based Safety Processes Since 1997" (break-out session at the 18th Annual Safety Performance Solutions Conference, Roanoke, VA, August 2014).

20. E. S. Geller, *The Psychology of Safety: How to Improve Behaviors and Attitudes on the Job* (Radnor, PA: Chilton Book Company, 1996); E. S. Geller, *Understanding Behavior-Based Safety: Step-by-Step Methods to Improve Your Workplace*, rev. ed. (Neenah, WI: J. J. Keller & Associates, Inc., 1998); T. E. McSween, *The Values-Based Safety Process: Improving Your Safety Culture with a Behavioral Approach* (New York: Van Nostrand Reinhold, 1995).
21. E. S. Geller, *People-Based Safety: The Source* (Virginia Beach, VA: Coastal Training and Technologies Corporation, 2005); E. S. Geller, "The Tragic Shooting at Virginia Tech: Personal Perspectives, Prospects, and Preventive Potentials," *Traumatology* (2008), 14(1):8–20.
22. E. S. Geller, *The Psychology of Safety: How to Improve Behaviors and Attitudes on the Job* (Radnor, PA: Chilton Book Company, 1996); E. S. Geller, *Understanding Behavior-Based Safety: Step-by-Step Methods to Improve Your Workplace,* rev. ed. (Neenah, WI: J. J. Keller & Associates, Inc., 1998); E. S. Geller, *The Psychology of Safety Handbook* (Boca Raton, FL: CRC Press, 2001).
23. E. S. Geller, "Why We Need Humanistic Behaviorism: Enhancing the Impact of Behavior-Based Safety," *Industrial Safety & Hygiene News* (2015), 49(10):52–53; E. S. Geller, S. Perdue, and A. French, "Behavior-Based Safety Coaching: Ten Guidelines for Successful Application," *Professional Safety* (2004), 49(7):42–49.
24. E. S. Geller, *The Psychology of Safety: How to Improve Behaviors and Attitudes on the Job* (Radnor, PA: Chilton Book Company, 1996.); E. S. Geller, *The Psychology of Safety Handbook* (Boca Raton, FL: CRC Press, 2001); E. S. Geller, "The Psychology of AC4P Behavior," ed. E. S. Geller, *Applied Psychology: Actively Caring for People* (New York: Cambridge University Press, 2016), pp. 45–82.
25. E. S. Geller, "Effective AC4P Communication," ed. E. S. Geller, *Applied Psychology: Actively Caring for People* (New York: Cambridge University Press, 2016), pp. 153–184.
26. D. Packard, *The HP Way: How Bill Hewlett and I Built Our Company* (New York: Harper Collins, 2006).
27. E. S. Geller, "The Psychology of Self-Motivation," ed. E. S. Geller, *Applied Psychology: Actively Caring for People* (New York: Cambridge University Press, 2016), pp. 83–118.
28. K. E. Weick, "Small Wins: Redefining the Scale of Social Problems," *American Psychologist* 39(1984):40–44.
29. E. L. Deci, *Intrinsic Motivation* (New York: Plenum, 1975); E. L. Deci and R. Flaste, *Why Do We Do What We Do: Understanding Self-Motivation* (New York: Penguin, 1995); E. L. Deci and R. M. Ryna, *Intrinsic Motivation and Self-Determinism in Human Behavior* (New York: Plenum, 1995); E. S. Geller, "The Psychology of Self-Motivation," ed. E. S. Geller, *Applied Psychology: Actively Caring for People* (New York: Cambridge University Press, 2016), pp. 83–118; E. S. Geller and B. Veazie, *The Motivation to Actively Care: How You Can Make It Happen* (New York: Morgan James Publishers, 2017).

30. D. Carnegie, *How to Win Friends and Influence People*, rev. ed. (New York: Simon and Schuster, 1981), p. 62.
31. D. A. Norman, *The Psychology of Everyday Things* (New York: Harper Collins Publishers, 1988).
32. Ibid; J. T. Reason and K. Mycieslka, *Absent Minded? The Psychology of Mental Lapses and Everyday Errors* (Englewood Cliffs, NJ: Prentice-Hall, 1982).
33. W. W. Heinrich, *Industrial Accident Prevention* (New York: McGraw-Hill, 1931).
34. F. E. Bird, Jr. and R. J. Davies, *Safety and the Bottom Line* (Loganville, GA: Institute Publishing, Inc., 1996).
35. A. Bandura, *Principles of Behavior Modification* (New York: Holt, Reinhold & Winston, 1969).
36. E. S. Geller, *The Psychology of Safety: How to Improve Behaviors and Attitudes on the Job* (Radnor, PA: Chilton Book Company, 1996); E. S. Geller, "Behavior-Based Safety in Industry: Realizing the Large-Scale Potential of Psychology to Promote Human Welfare," *Applied & Preventive Psychology* 10(2001):87–105; E. S. Geller, *The Psychology of Safety Handbook* (Boca Raton, FL: CRC Press, 2001); ed. E. S. Geller and J. H. Williams, *Keys to Behavior-Based Safety* (Rockville, MD: ABS Consulting, 2001); D. Petersen, *Safe Behavior Reinforcement* (Goshen, NY: Aloray, Inc., 1989); B. Sulzer-Azaroff and J. Austin, "Does BBS Work? Behavior-Based Safety and Injury Reduction: A Survey of the Evidence," *Professional Safety* (2000), 45(7):19–24.
37. E. S. Geller, *Understanding Behavior-Based Safety: Step-by-Step Methods to Improve Your Workplace*, 2nd ed. (Neehah, WI: J. J. Keller & Associates, Inc., 1998); E. S. Geller, *Beyond Safety Accountability* (Rockville, MD: Government Institutes, 2001); E. S. Geller, *The Psychology of Safety Handbook* (Boca Raton, FL: CRC Press, 2001); E. S. Geller, *Working Safe: How to Help People Actively Care for Health and Safety*, 2nd ed. (New York: Lewis Publishers, 2001); E. S. Geller, "Behavior-Based Safety for Construction Safety: From Research-Supported Principles to Practical Procedures," Proceedings of Construction Safety Leadership Symposium (Des Plaines, IL: American Society of Safety Engineers, 2003); T. E. McSween, *The Value-Based Safety Process: Improving Your Safety Culture with a Behavioral Approach* (New York: Van Nostrand Reinhold, 1995).
38. K. Blanchard, *Building Gung Ho Teams: How to Turn People Power into Profits* (workshop presented at Hotel Roanoke, Roanoke, VA, November 1999); W. C. Byham, *Zapp! The Lightning of Empowerment* (New York: Fawcett Columbine, 1998).
39. D. L. Watson and R. G. Tharp, *Self-Directed Behavior: Self-Modification for Personal Adjustment*, 8th ed. (Pacific Grove, CA: Brooks/Cole Publishing, 2002).
40. E. S. Geller, *Beyond Safety Accountability* (Rockville, MD: Government Institutes, 2001).
41. R. A. Monty et al., "The Freedom to Choose Is Not Always So Choice," *Journal of Experimental Psychology: Human Learning and Memory* 37(1979):170–178;

L. C. Perlmutter, R. A. Monty, and G. A. Kimble, "Effect of Choice on Paired-Associate Learning," *Journal of Experimental Psychology* 91(1971):47–53.

42. J. P. DePasquale and E. S. Geller, "Critical Success Factors for Behavior-Based Safety: A Study of Twenty Industry-wide Applications," *Journal of Safety Research* (1999), 30(4):237–249.
43. E. S. Geller et al., "Researching Behavior-Based Safety: A Multi-Method Assessment and Evaluation," Proceedings of the 37th Annual Professional Development Conference and Exposition (Des Plaines, IL: American Society of Safety Engineers, June, 1998), pp. 537–559.
44. T. R. Krause, J. H. Hidley, and S. J. Hodson, *The Behavior-Based Safety Process: Managing Improvement for an Injury-Free Culture*, 2nd ed. (New York: Van Nostrand Reinhold, 1996).
45. D. Grote, *Discipline Without Punishment* (New York: American Management Association, 1995); E. S. Geller, *The Psychology of Safety: How to Improve Behaviors and Attitudes on the Job* (Radnor, PA: Chilton Book Company, 1996); E. S. Geller, *The Psychology of Safety Handbook* (Boca Raton, FL: CRC Press, 2001); E. S. Geller, *Working Safe: How to Help People Actively Care for Health and Safety*, 2nd ed. (New York: Lewis Publishers, 2001); M. Sidman, *Coercion and Its Fallout* (Boston, MA: Authors Cooperative, 1989).
46. C. R. Rogers, *Client-Centered Therapy* (Boston, MA: Houghton-Mifflin, 1951).
47. A. C. Daniels, *Performance Management*, 3rd ed. (Tucker, GA: Performance Management Publications, 1989); E. S. Geller, *The Psychology of Safety Handbook* (Boca Raton, FL: CRC Press, 2001); E. S. Geller and Z. J. Robinson, "Behavior-Based Approaches to Occupational Safety," ed. S. Clarke et al., *The Psychology of Occupational Safety and Workplace Health* (New York: Wiley, 2016), pp. 301–326.
48. F. A. Manuele, "Perspectives on Behavioral Safety: Observations of ASSE's Behavior Safety Symposium," *Professional Safety* 43(August, 1998):32–37; T. A. Smith, "Viewpoint: Rebutting Behaviorism," *Industrial Safety & Hygiene News* 40(1995):40; R. M. Yandrick, *Behavioral Risk Management* (San Francisco, CA: Jossey-Bass Publishers, 1996).
49. M. Hans, "Does Behavior-Based Safety Work?" *Safety and Health* (National Safety Council, June, 1996), pp. 44–49; T. R. Krause, J. H. Hidley, and S. J. Hodson, *The Behavior-Based Safety Process: Managing Improvement for an Injury-Free Culture*, 2nd ed. (New York: Van Nostrand Reinhold, 1996); D. Petersen, "Behavior-Based Safety: Magic Pill, Fad, or What?" *Professional Safety* 43(1998):34.
50. E. S. Geller, *The Psychology of Safety Handbook* (Boca Raton, FL: CRC Press, 2001); E. S. Geller, *Working Safe: How to Help People Actively Care for Health and Safety*, 2nd ed. (New York: Lewis Publishers, 2001); E. S. Geller, "Behavior-Based Safety for Construction Safety: From Research-Supported Principles to Practical Procedures," Proceedings of Construction Safety Leadership Sympo-

sium (Des Plaines, IL: American Society of Safety Engineers, 2003); ed. E. S. Geller and J. H. Williams, *Keys to Behavior-Based Safety* (Rockville, MD: ABS Consulting, 2001).

51. C. S. Dweck, *Mindset: The New Psychology of Success* (New York: Ballantine Books, 2006.).
52. P. Chance, *The Teacher's Craft: The 10 Essential Skills of Effective Teaching* (Long Grove, IL: Waveland Press, Inc., 2008); D. Reed et al., "Actively Caring for Higher Education," ed. E. S. Geller, *Applied Psychology: Actively Caring for People* (New York: Cambridge University Press, 2016), pp. 563–593.
53. E. L. Thorndike, *Human Learning* (Cambridge, MA: MIT Press, 1931).
54. E. S. Geller, *The Psychology of Safety Handbook* (Boca Raton, FL: CRC Press, 2001); E. S. Geller, "The Psychology of AC4P Behavior," ed. E. S. Geller, *Applied Psychology: Actively Caring for People* (New York: Cambridge University Press, 2016), pp. 45–82.
55. D. Carnegie, *How to Win Friends and Influence People*, rev. ed. (New York: Simon and Schuster, 1981), p. 48.
56. D. Carnegie, *How to Win Friends and Influence People*, rev. ed. (New York: Simon and Schuster, 1981), p. 130.
57. *New Merriam Webster Dictionary*, s.v. "sympathy."
58. *New Merriam Webster Dictionary*, s.v. "empathy."
59. *The American Heritage Dictionary*, Second College Edition, s.v. "empathy."
60. *The American Heritage Dictionary*, Second College Edition, s.v. "sympathy."
61. S. R. Covey, *The Seven Habits of Highly Effective People* (New York: Simon and Schuster, 1989), pp. 236–260.
62. W. E. Deming, "Quality, Productivity, and Competitive Position" (workshop presented by Quality Enhancement Seminars, Inc., Cincinnati, OH, May 1991).
63. A. H. Maslow, "A Theory of Human Motivation," *Psychological Review* 50(1943): 370–396; A. H. Maslow, *Motivation and Personality* (New York: Harper, 1954).
64. A. H. Maslow, *The Farther Reaches of Human Nature* (New York: Viking, 1971).
65. R. B. Cialdini, *Influence: Science and Practice*, 4th ed. (Needham Heights, MA: Allyn & Bacon, 2001); C. Furrow and E. S. Geller, "Social Influence and AC4P Behavior," ed. E. S. Geller, *Applied Psychology: Actively Caring for People* (New York: Cambridge University Press, 2016), pp. 185–227.

# GLOSSARY OF KEY TERMS

**ABC (activator-behavior-consequence):** The three-term contingency of applied behavioral science, specifying that stimulus events preceding behavior provide direction while the consequences following behavior provide motivation for the recurrence of that behavior.

**accountability system:** An extrinsic contingency or intervention that motivates people to perform certain behavior in order to obtain a positive consequence or avoid a negative consequence.

**activator:** An environmental event (e.g., direction from a sign or individual) implemented to influence the occurrence of a particular behavior.

**actively caring:** The combination of action and compassion, resulting in behavior performed on behalf of the safety, health, and/or well-being of one or more other persons.

**actively caring for people (AC4P):** The application of behavioral science and select principles from humanism in order to increase the frequency and improve the quality of behavior that benefits human safety, health, and/or welfare—referred to as humanistic behaviorism.

**applied behavioral science (ABS):** Application of research-based principles derived from experimental and applied behavior analysis to increase the occurrence of desirable behaviors and decrease the frequency of undesirable behaviors.

**AC4P coaching:** Interpersonal communication whereby one individual (the teacher) employs principles of humanistic behaviorism to benefit the observed behavior of another individual (the learner).

**AC4P culture:** An environmental context or setting (e.g., a workplace, school, or home) where people interact daily on behalf of the safety, health, and well-being of everyone else in the surroundings with a spirit of interdependence and self-transcendence.

**AC4P movement:** Principle-centered activities of individuals on a mission to increase the large-scale occurrence of AC4P behaviors throughout organizations, schools, homes, and communities at large.

**AC4P wristband:** A green silicone wristband engraved with "Actively Caring for People" and used to reward acts of kindness and promote participation in the AC4P movement.

**behavior-based safety (BBS):** The systematic application of ABS to increase the occurrence of safe behavior and decrease the frequency of at-risk (or unsafe) behavior, typically manifested by a peer-to-peer behavioral observation and feedback process of interpersonal coaching.

**behavior-based feedback:** Interpersonal communication following the occurrence of a behavior that informs the performer what he or she did correctly (supportive feedback) and/or incorrectly (corrective feedback).

**belongingness:** Perceived mutual social support or interdependency with others—friends, family, colleagues, or even strangers.

**brother's/sister's keeper culture:** An environmental setting (e.g., workplace, school, or home) in which all participants feel empowered and self-motivated to routinely perform AC4P behavior on behalf of the safety, health, and well-being of everyone else in that setting.

**choice:** The perception of having more than one option with regard to accomplishing a particular task or action plan.

**cognitive failure:** An unintentional error or mistake caused by a lack of appropriate attention or thinking (self-talk) about ongoing behavior.

**community:** The perception of having interdependent social or group support regarding the accomplishment of a certain challenge or assignment.

**competence:** The perception or belief that an individual has the knowledge, skill, and ability to accomplish a certain task effectively.

**conscious competence:** The performer of desirable (e.g., safe) behavior is acting intentionally or mindfully, perhaps with an AC4P mind-set.

**conscious incompetence:** The performer of undesirable (e.g., unsafe) behavior is aware of the risks and/or the rule or regulation disallowing that behavior.

**consequence:** An environmental event (positive or negative) that naturally occurs (i.e., intrinsic) after a behavior or is added to the situation (i.e., extrinsic) to influence a certain behavior, which may or may not influence the frequency or form of the preceding behavior.

**contingency:** An if/then relationship between a particular behavior and a certain consequence—"*If* you do that, *then* you will receive this consequence."

**corrective feedback:** Behavior-based information given to an individual after observation of an undesirable (e.g., at-risk) or less-than-optimal behavior, often manifested in a one-on-one coaching interaction.

**critical behavior checklist (CBC):** A list of safe and at-risk behaviors defined by a work team and used to systematically record the occurrence of safe and at-risk behaviors during a peer-to-peer behavioral-observation-and-feedback session.

**discipline:** A behavior-based intervention designed to teach and/or motivate an individual to perform a particular target behavior in a certain way—derived from the Latin word *disciplina*, which means "instruction."

**discretionary behavior:** Desirable behavior that was not required by an extrinsic accountability system—presumed to be self-directed and self-motivated.

**DO IT process:** An acronym reflecting the basic scientific method of applied behavioral science (ABS): *D* for Define the target behavior to influence, *O* for Observe the natural occurrence frequency of the target behavior (i.e., baseline), *I* for Intervene to change the frequency of the target behavior in a desirable direction (i.e., increase or decrease), and *T* for Test the impact of the intervention by observing the target behavior during and after the intervention ends and comparing these frequencies of occurrence with the baseline frequency.

**education:** Teaching one or more persons awareness, theory, or procedures related to a certain task or the circumstances related to that task.

**emotion:** A potent and potentially long-term reaction to a person or situation that can motivate relevant self-directed behavior.

**empathic listening:** The highest level of interpersonal listening whereby the listener attempts to identify with and understand the speaker's feelings, attitude, motives and behavior, as well as the circumstances that influenced relevant person-states and behavior.

**empathy:** Attempting to identify with and understand another individual's circumstances that influence his or her attitude, feelings, motives, and behavior.

**empowerment:** An individual's belief or confidence that he or she can master a particular assignment or accomplish a certain task—determined by answering yes to three questions: Can I do it? Will it work? Is it worth it?

**extrinsic:** A stimulus event external to the individual, considered to be an activator or consequence when applied to influence a target behavior.

**extrinsic stimulus:** An observable and external environmental event that might direct behavior (as an activator) or motivate behavior (as a positive or negative consequence).

**failure accepter:** An individual who has surrendered all possibilities of successfully avoiding a particular negative consequence—submitting to a person-state of helplessness.

**failure avoider:** An individual who performs a certain behavior in order to avoid a negative consequence.

**feedback:** Information following the performance of a behavior that might influence the frequency and/or form of that preceding behavior by supporting or correcting it.

**feedforward:** Information preceding the performance of a behavior (e.g., directions, expectations, mandates) that might activate behavioral occurrence, perhaps influencing the frequency and/or the form of the following behavior.

**goal setting:** The systematic process of defining a behavior-focused SMARTS goal.

**humanism:** An appreciation of empathy, diversity, compassion, and human dignity—a sincere concern for human welfare and well-being.

**humanistic behaviorism:** Making evidence-based interventions from applied behavioral science (ABS) more effective by practicing empathy and considering such person-states as self-esteem, self-efficacy, belongingness, personal control, and optimism.

**incentive:** An announcement (an activator) that specifies the occurrence of a positive consequence (a reward) following performance of a designated behavior or a desirable outcome of more than one behavior.

**incompetence:** The perception or belief that an individual lacks the knowledge, skill, and/or ability to perform effectively the behaviors required to complete a certain task.

**independence:** An individualistic viewpoint or mind-set that sets the occasion for self-focused behavior ("I can do this myself"), and inhibits other-focused behavior ("I don't need your help").

**interdependence:** A collectivistic, we-need-each-other viewpoint or mind-set that sets the occasion for AC4P behavior—acts performed on behalf of the safety, health, or well-being of others.

**intervention:** An external or extrinsic program or process implemented to influence the quantity or quality of one or more target behaviors.

**intrinsic:** A behavioral consequence that occurs naturally and motivates the recurrence of the behavior as a reinforcer.

**leadership:** Inspiring people to be self-motivated to perform one or more goal-directed and discretionary behaviors.

**management:** Motivating one or more behaviors to occur through an external or extrinsic accountability system (e.g., activator, behavioral feedback, incentive, disincentive, rule, command, interpersonal recognition or reprimand).

**nondirective coaching:** Empathic listening occurs to learn another person's perceptions and circumstances before advice or direction is offered.

**observational learning:** Learning how to perform one or more behaviors by watching one or more individuals perform the behavior(s).

**optimism:** An expectation that the consequences of one or more behaviors will be positive—the challenge will be met successfully.

**other-directed behavior:** Behavior that occurs as a result of an extrinsic or external intervention, such as an incentive that promises a reward following a target behavior or a disincentive that warns the levy of a penalty if a certain undesirable behavior occurs.

**outcome expectancy:** An individual's belief or confidence that the anticipated consequences following the performance of one or more goal-directed behaviors is worth the effort expended and the time involved to achieve the desired outcome.

**participative management:** Empowering people in the design and/or implementation of the contingencies or the accountability system that controls one or more of their behaviors.

**penalty:** A negative consequence following one or more behaviors that may or may not influence subsequent performance.

**percent-safe score:** The percentage of safe behaviors observed by systematically completing a critical behavior checklist (CBC) during a peer-to-peer behavioral-observation-and-feedback session, calculated by dividing the total number of safe behaviors observed by the total number of behavioral observations (safe plus at-risk behaviors) and multiplying the quotient by 100.

**personal control:** A perception of ability or competence to accomplish a particular task or handle a certain challenge.

**person-state:** A personal disposition (e.g., expectancy, attitude, or personality factor) that influences an individual's behavior.

**pessimism:** An expectation that particular behavior or a certain action plan will not be successful or will result in a negative consequence, thereby reflecting Murphy's Law: "If anything can go wrong, it will."

**principle of consistency:** The research-based dynamic that people will alter their attitudes, beliefs, and perceptions to be consistent with their behavior, and vice versa.

**principle of reciprocity:** A research-based social norm supporting the notion of payback or "pay it forward," an obligation to return a good turn (a favor) or a bad turn (a negative interaction).

**punisher:** A negative consequence following a designated behavior that reduces the occurrence of that behavior.

**punishment:** A behavior-focused intervention that successfully reduces the frequency of a target behavior by following that behavior with a negative consequence.

**reciprocity norm:** The social norm that obligates people to repay others with the form of behavior (positive or negative) they had received from them.

**response-efficacy:** An individual's belief or confidence that performing one or more goal-directed behaviors will contribute to achieving a desirable outcome or long-term mission.

**reward:** A positive consequence following a designated behavior or an outcome of more than one behavior that may or may not influence subsequent performance.

**safety incentive program:** A common approach to motivating engagement in safety by offering participants a reward if the number of reported injuries decreases to a certain level. The result: a decrease in reported injuries but minimal impact on safety-related behavior.

**SAPS (See, Act, Pass, Share):** The four-step process for delivering an AC4P wristband to reward AC4P behavior and promote the AC4P movement: See an AC4P behavior, Act to reward the AC4P behavior, ask the wristband recipient to Pass on the wristband to reward another person's AC4P behavior, and Share the positive AC4P interaction at the ac4p.org website.

**selection by consequences:** B.F. Skinner's most important legacy that emulates the motivational principle that people do what they do to gain a positive consequence or to avoid or escape a negative consequence.

**self-accountability:** Internal motivation (from within) to perform one or more behaviors.

**self-actualization:** The top of Maslow's *initial* hierarchy of needs at which a person feels a sense of ultimate achievement by fulfilling one's potential.

**self-directed behavior:** Self-motivated behavior that is not activated or motivated by an external or extrinsic intervention (e.g., an incentive or disincentive).

**self-efficacy:** An individual's belief or confidence that he or she has the knowledge, skill, and ability to perform the goal-directed behaviors needed to accomplish a certain task.

**self-esteem:** A general or overall feeling of self-worth that influences one's propensity to perform AC4P behavior.

**self-motivation:** A person-state that reflects internal drive to perform certain behavior or achieve a particular outcome of one or more behaviors from a self-directed mind-set.

**self-transcendence:** The top of Maslow's *revised* hierarchy of needs whereby an AC4P mind-set is realized, and the individual experiences personal fulfillment and reinforcement when performing AC4P behavior—behavior contributing to the safety, health, or welfare of another person.

**SMARTS goal:** A behavior-focused achievement objective defined by making it Specific, Motivational, Attainable, Relevant, Trackable, and Shared.

**success seeker:** An individual who performs one or more behaviors in order to earn a positive consequence, which can be intrinsic or extrinsic.

**synergy:** Interdependent participation in a goal-directed activity that results in greater achievement than would be possible if all participants worked independently toward the same goal.

**total recordable injury rate (TRIR):** A measure of the rate of recordable workplace injuries, normalized per 100 workers per year—derived by multiplying the number of recordable injuries in a calendar year by 200,000 (100 employees working 2,000 hours per year) and dividing this value by the total hours employees actually worked that year.

**training:** Providing behavior-based direction or feedback regarding the performance of one or more specified behaviors relevant to achieving a designated outcome.

**unconscious competence:** The performer of desirable (e.g., safe) behavior is mindless or unintentional with regard to that behavior—typically referred to as desirable habitual behavior.

**unconscious incompetence:** The performer of undesirable (e.g., at-risk) behavior is unaware or mindless with regard to that behavior—referred to as desirable habitual behavior.

**vicarious punishment:** Observational learning whereby the negative consequence given to one or more persons following an undesirable (e.g., at-risk) behavior influences an observer of this interaction to decrease his or her performance of that behavior.

**vicarious reinforcement:** Observational learning whereby the positive consequence given to one or more persons following a desirable (e.g., safe) behavior influences an observer of this interaction to increase the frequency or improve the quality of his or her performance of that behavior.

**vision:** A long-term or ultimate objective or outcome achieved through serial goal-setting and successive realization of these behavior-based goals.

# SUGGESTED READING

Biglan, A. *The Nurture Effect: How the Science of Human Behavior Can Improve Our Lives and Our World*. Oakland, CA: New Harbinger Publications, Inc., 2015.

Covey, S. R. *The Seven Habits of Highly Effective People: Restoring the Character Ethic*. New York: Simon & Schuster, 1989.

Daniels, A. C. *Bringing Out the Best in People: How to Apply the Astonishing Power of Positive Reinforcement*. New York: McGraw-Hill, Inc., 1994.

Drebinger, J. W. *Mastering Safety Communication: Communication Skills for a Safe, Productive and Profitable Workplace*. Galt, CA: Wulamoc Publishing, 1997.

Drebinger, J. W. *Would You Watch Out for My Safety? Helping Others Avoid Personal Injury*. Galt, CA: Wulamoc Publishing, 2011.

Geller, E. S., ed. *Applied Psychology: Actively Caring for People*. New York: Cambridge University Press, 2016.

Geller, E. S. *The Participation Factor: How to Increase Involvement in Occupational Safety*. Des Plaines, IL: American Society of Safety Engineers, 2002.

Geller, E. S., and B. Veazie. *The Courage to Actively Care: How to Cultivate a Culture of Interpersonal Compassion*. New York: Morgan James Publishers, 2017.

Geller, E. S., and B. Veazie. *The Motivation to Actively Care: How You Can Make It Happen*. New York: Morgan James Publishers, 2017.

Norman, D. A. *The Psychology of Everyday Things*. New York: Basic Books, 1988.

Petersen, D. *Safety Management: A Human Approach*. 3rd ed. Des Plaines, IL: American Society of Safety Engineers, 2001.

Watson, D. L., and R. G. Tharp. *Self-Directed Behavior: Self-Modification for Personal Adjustment*. 8th ed. Belmont, CA: Thompson Learning, Inc., 2002.

Williams, J. H. *Keeping People Safe: The Human Dynamics of Injury Prevention*. Lanham, MD, Government Institutes/Scarecrow Press, 2010.

# ABOUT THE AUTHORS

**E. Scott Geller**, Ph.D., an Alumni Distinguished Professor at Virginia Tech, is cofounder and senior partner of Safety Performance Solutions, Inc. (safetyperformance.com), a leading-edge training and consulting organization specializing in AC4P safety since 1995. For almost five decades, Professor Geller has taught and conducted research as a faculty member and director of the Center for Applied Behavior Systems in the Department of Psychology at Virginia Tech.

He has authored, edited, or coauthored 37 books, 82 book chapters, 39 training programs, 259 magazine articles, and more than 300 research articles addressing the development and evaluation of behavior change interventions to improve quality of life on a large scale. His most recent textbook, *Applied Psychology: Actively Caring for People*, defines Dr. Geller's research, teaching, and scholarship career at Virginia Tech, which epitomizes the VT motto *Ut Prosim* (That I May Serve).

His popular books in applied psychology include: *The Psychology of Safety: Improving Behaviors and Attitudes on the Job; Working Safe; Understanding Behavior-Based Safety; Building Successful Safety Teams; Beyond Safety Accountability: How to Increase Personal Responsibility; The Psychology of Safety Handbook; Keys to Behavior-Based Safety from Safety Performance Solutions; The Participation Factor: How to Increase Participation in Occupational Safety; People-Based Safety: The Source; People-Based Patient Safety: Enriching Your Culture to Prevent Medical Error* (coauthored with Dave Johnson); *Leading People-Based Safety: Enriching Your Culture;*

*Actively Caring for People: Cultivating a Culture of Compassion; The Courage to Actively Care: How to Cultivate a Culture of Interpersonal Compassion* (coauthored with Bob Veazie); *The Motivation to Actively Care: How You Can Make It Happen* (coauthored with Bob Veazie); *Actively Caring for People Policing: Building Positive Police/Citizen Relations* (coauthored by Bobby Kipper); and *Actively Caring for People in Schools: How to Cultivate a Culture of Compassion.*

Dr. Geller is a fellow of the American Psychological Society, the Association for Psychological Science, the Association of Behavior Analysis International, and the World Academy of Productivity and Quality Sciences. He is a past editor of the *Journal of Applied Behavior Analysis* (1989–1992); current associate editor of *Environment and Behavior* (since 1982); and current consulting editor for *Behavior and Social Issues*, the *Journal of Organizational Behavior Management*, and the *Journal of Safety Research*.

He has received lifetime achievement awards from the International Organizational Behavior Management Network (2008) and the American Psychological Foundation (2009). And in 2010, he was honored with the Outstanding Applied Research Award from the American Psychological Association's Division of Applied Behavior Analysis. In 2011, the College of Wooster awarded E. Scott Geller the honorary degree Doctor of Humane Letters.

**Krista S. Geller**, Ph.D., has been involved in applied behavioral science since childhood as a subject in many of her father's psychology studies. With no informed consent or opportunity to opt out, Krista made the best of these "forced" learning experiences. It wasn't until her freshman year at Radford University that she realized the critical importance and social validity of her dad's research. It was then she decided to pursue a double major in psychology and media studies.

As a *New River Valley News* reporter for Radford University, many of her news stories related to her dad's research topics, but this time she was a willing participant. She produced such media stories as: the proper installation of child safety seats, strategies for healthy eating, the multiple effects of coal mines in Virginia, the creation of the "road rage reducer," the impact of having a date on the alcohol consumption of college students, and the significance of the Actively Caring for People (AC4P) movement." Over the years, Krista had the unique opportunity to voluntarily participate in these and many other research studies at Virginia Tech's Center for Applied Behavior Systems (CABS), where she worked as a researcher throughout her undergraduate and graduate studies.

In 2002 and 2005, Krista earned her master's and doctorate degrees, respectively, in human development at Virginia Tech. Her focus was in Family Studies, Gerontology, and Marriage and Family Therapy, pinpointing the significance of interpersonal relationships throughout the human lifespan. Her exposure to human-relations research led Krista to compare systematically the aging populations in Sweden versus the United States. Later, Krista's research interests shifted to her lifelong love of pets and the special impact of pet-people relationships on human dynamics. This research was highlighted in her recent book chapter, "The Power of Pets," published in the textbook *Applied Psychology: Actively Caring for People* (Cambridge University Press, 2016).

Krista spent five years teaching college students at Radford University, Virginia Tech, and Kingwood College in Texas. Her passion to keep people safe returned when Krista received opportunities to work as an ESH representative for several companies. In 2012, she became the People-Based Safety (PBS) Global Manager for Bechtel Corporation. As a PBS teacher, coach, and consultant for Bechtel, Krista helped employees develop the competence and courage to actively care for the safety of themselves and their coworkers. She has travelled worldwide to assist in adapting the PBS process to various cultures, languages, and work processes. Her mission is to inspire people to take the time to actively care for the safety of themselves and others. She is now working as the Field Leadership Development Supervisor, Bechtel Corporation.

# INDEX

# NOTES

# NOTES

# NOTES

# NOTES

# NOTES